Customer Relations and Sales From A to Z

For Aerospace, Defense and Technology Companies Doing Business with Governments

David E. Potts, Ph.D.

Solving the Sales Puzzle ®

Copyright © 2015 by Tazewell Strategies. All rights reserved. Except as permitted under the United States copyright Act of 1976, no part of this publication may be reproduced or distributed in any form by any means, or stored in a database retrieval system, without the prior permission of the author.

ISBN: 978-0-9861495-0-4
Library of Congress Control Number: 2015942686

Dedication

To my father, George, who revealed the value of investing in sales relationships

To my mother, Vivian, who showed me how to relate to difficult customers

To my sister, Judy, who teaches me the honor of serving others

To my wife, Paula, who cautions me to stop talking and start listening

To my son, Greg, who reminds me that now is always the right time to do the right thing

Table of Contents

Introduction 1
Be an Ace 5
Alternatives 10
Authenticity 12
Baseball and Business 17
Buying 22
Challenge Your Customer 27
Coaches 32
Conversation 36
Create the Customer 40
Customer-Centric Culture 45
Differentiation in Competitions 50
Day in the Life of a Customer 55
Emotions 60
Emotional Decisions 64
Fear 69
Find a Problem, Fix a Problem 73
Focus on the Customer 78
Generations 83
Golf and Life and Business 89
Handling Difficult Customers 93
Handshakes 99
Selling to Humans 102
Integrity 108
Customer Intimacy 112
Jolt Your PowerPoints and Presentations 118
Keep It Short 129
Killer Argument 132
Lessons Learned 136
Listening 140
Make *a Smile* 146
Sales Management 150
Saying No 156
Objections to Price 161
Opportunities 164
Pain 170
People 174
Personalities 178
Sales Process 182

Promises	187
Why Customers **Q**uit	191
Questions	196
Customer **R**elations	202
Risk	206
Sales	212
Service	216
Strategy and Tactics	220
Telephones	225
Teams	229
Trust	234
Understanding Buyer-Seller Alignment	239
Voice of the Customer	244
Voice Tone	247
Videos	249
Words	256
Wise Words	259
X-Ray Vision	264
Your Call to Action	269
*Make It E-***Z**	274
Conclusion	279
Suggested Reading	281
Customer Relations and Sales Training Partners	285

Introduction

What follows in this book are some of the "discoveries" I've made over the past decade while teaching customer relations classes. There are no industrial secrets here. No proprietary information. No singular list of best practices. But, hopefully, there will be revelations for those who want to do a better job serving their government customers. I've arranged the book in alphabetical order because I haven't yet come up with a universal theory for government customer relations and sales. Recommendations for training partners and great books on the subject can be found in back. And I've included lots of research references and sets of quotes for each subject, as I've found that engineer-based corporate cultures disdain "soft-skill" techniques not grounded in data. Plus, everyone appreciates a good quote giving greater perspective. The quotes generally support and reinforce the subjects, though I have to admit some are inserted just because I like them.

Over the years I've given a lot of thought to how people in the business of selling to government customers go about their work. When I was hired into business development at the Lockheed Martin Aeronautics Company I thought I knew what I was doing. I had spent the first years of my Air Force career working with Boeing and TRW contractors on the Minuteman ICBM system in Cheyenne, Wyoming. Later at the Pentagon I worked on the National Disclosure Policy Committee coordinating Air Force international defense sales. In Korea I worked with Raytheon on information and intelligence networks. And when I was the Defense Attaché in Prague in the late 1990s, I worked with every major defense contractor assisting the Czechs in getting ready to join NATO. So when I entered the business world, I was certain I knew how the business development game was played. But I didn't.

It took me years to begin to understand relationships between the aerospace, defense, and technology companies and their domestic and international customers. I learned that government customers are administrative customers buying for others, not consumer customers buying for themselves. Their motives and behaviors can be quite different from examples used in academic case studies and commercial sales courses. And yet, because government customers are human beings, they still adhere to age-old traditions and traits of buying and selling. Along the way I also recognized that sales was in my DNA: my great-grandfather sold cattle, my grandfather sold grain, my father sold paint and I had been selling myself and my ideas all through my life, even though I spent a large part of it in military service. I was a salesman. I just hadn't acknowledged it.

After I secured my new position at the Lockheed Martin Aeronautics Company, returning to Fort Worth after many years absence, I proudly told my Dad that I had a business development job. He said, "Great, you're in sales." I bristled a little and corrected him. He laughed and asked me to tell him what I'd be doing. I explained all about shaping the business environment and capturing business with our processes and procedures which had been honed over a quarter of a century of fierce and unrelenting competition for fighter and transport aircraft contracts in the U.S. and abroad. Dad nodded, "Like I said, you're in sales."

Business developers in aerospace, defense and technology companies don't think of themselves as salespeople, do they? Many arrive at business development positions either coming up from programs or coming over from the customer base. Former program managers tend to sell only from a capability-cost-schedule perspective. Former customers tend to sell only from a glad-handing perspective. Both can be good, but neither is sufficient. In my early years in the airplane business I was astonished at how some of our superstar business developers could put a multimillion-dollar deal together and I tried to emulate them. But it was hit-and-miss. The superstars generally were not good mentors, for they operated more on gut feel than standard procedures and had little patience teaching others. They couldn't fully explain why they won; and when they lost, it always was the customer's fault.

I did well enough in my business development work that after a few years I gained a headquarters position in a small best practices section dedicated to analyzing and sharing information on what we do right when we win and what we do wrong when we lose. During my interview for the job I was asked if I could put together a customer relations course to help our business development community. I had no idea if I could do this, but because it was an interview and I wanted the job I said yes before my rational brain took charge. That snap decision led to some of the most fun and rewarding work I've ever had and lasting relationships with people like Burlie Brunson, George Burgess and Chip Schleider inside Lockheed Martin, and my outside mentor and co-instructor Dr. Tom Barrett, President of Business/Life Management.

I sincerely hope you will find this information helpful in dealing with your government customers. They have difficult jobs and you don't want to make it any harder for them. Aerospace, defense, and technology companies supporting governments do important work. I wish each and every reader the best of success in providing superior quality products and services at reasonable prices at the right times to help their customers support more effective and efficient governance. Now to finish with a quote from my former boss and defense industry icon Bob Trice, "Please go out there and sell something."

BE AN ACE - ALTERNATIVES - AUTHENTICITY

Be an Ace

It's a fact that you can't be good at customer relations and sales if you're not good at speaking and presenting. To be clear and compelling is an art as much as a technique and you must deliberately apply yourself. Do you know people who speak easily and are always talking? Some of them are effective communicators and some are just talkers. I knew a Hungarian diplomat who could speak five languages - but had nothing interesting to say in any of them. When I was young I was more the opposite, I believed I had interesting things to say, but the words came out too fast and jumbled up.

In my first semester at college I was an international affairs major. International studies eventually became my ticket to success in the military and business, but as a seventeen-year-old I was uninspired by old guys reading lectures from yellowed notes. Art De Rojas, a guitar-playing Cuban-American in the fraternity I was pledging advised me to switch majors to Radio-TV-Film, because it was fun and there were lots of girls in it. It seemed counter-intuitive, but that was one of the best decisions I ever made. It opened me up to the worlds of music, art, philosophy and all types of communication skills which continue to help me to this day.

The Communications Department at Texas Christian University in those days was run by Dr. R.C. Norris and Professor Larry Lauer. They were consummate professionals and prepared us well, especially for the broadcasting industry. I didn't go on to work at radio and television stations like my peers, but I profited from being with them under the tutelage of Norris and Lauer. The class that scared me the most was public speaking. I made so many mistakes in the beginning and was always worried about what people were thinking of me. Over time, I learned that if I prepared well, slowed down and focused on the audience rather than myself I was much, much better.

Growing up in Fort Worth and living in a neighborhood of General Dynamics and Air Force families, I became fascinated with airplanes and dreamed of flying fighters. I didn't pass the flying physical, but went into the Air Force anyway and had a wonderful career being around some remarkable aircraft and some remarkable flyers. I had lunch once with aviation pioneer Ira Eaker, spent an evening with Congressional Medal of Honor winner George "Bud" Day, and worked alongside former Thunderbirds leader Jim Latham. I even spent a weekend in Vermont on a speaking tour with German WWII ace Adolf Galland.

There's something telling about people who master their professions, who apply their skills while controlling their emotions. Like aviation heroes, the good ones are not boastful, but they are not quiet, either. For many years I've taught customer relations classes together with Dr. Tom Barrett, President of Business/Life Management. Tom has worked with CEOs, members of congress and general officers, helping them improve their leadership and communication skills. One of his best-ever pieces of advice for leaders to improve their communication skills is to be an "ACE." He recalled meeting Joe Foss, a man who seemed to have crammed five lifetimes into the span of one. Joe was the leading U.S. Marine Corps Ace in World War II and later a general in the Air National Guard, governor of South Dakota, first commissioner of the American Football League, and finally a television broadcaster. Most of us would settle for just one of those accomplishments.

Dr. Barrett often refers to Joe Foss in his presentations on leadership and communication and I find his advice particularly useful for any customer relations or sales representative seeking better communication with customers and audiences. Without a good sense of leadership, you can't successfully manage a customer relationship or help a buyer make a decision. When Dr. Barrett encourages us to become an ACE, he means the type of person who consistently displays three qualities during interactions. Being an ACE means speaking with Authority, Conviction and Enthusiasm.

Authority - real authority - comes from you as a person more than your title. Even though it's an internal mindset, others can detect it immediately. You show that you're comfortable, confident and competent.

It becomes self-evident to your audience that you're up to the task at hand and it motivates people to trust you and listen to you.

Conviction - is likewise an external display of an internal belief that you're on the right path. It's the kind of certitude that keeps you and others moving forward in the face of obstacles. You telegraph positivity that what you're saying will lead to success in the end and that there are no walls that can hold you back.

Enthusiasm - can be a huge asset if rooted in sound thinking, clear vision and a repertoire of experience. But it can be a huge liability if it's only hype, smoke and mirrors. The less substantive and more enthusiastic you appear, the more your customers and audience become cynical. But genuine enthusiasm is infectious and creates positive energy.

It's interesting to observe that these three must go together. If you speak with authority, but without conviction and enthusiasm, you're boring. If you speak with conviction, but without authority and enthusiasm, you're not convincing. And if you speak with enthusiasm, but without authority and conviction, you're ignored. The ACE model means that you know what you're talking about, you believe in what you're talking about, and you are excited about what you're talking about. It's a powerful technique and emulates the behaviors of real aces. Here are quotes from some famous flyers to help get you in the right frame of mind to attack your next presentation and score a victory.

Ace Quotes

"A squadron commander who sits in his tent and gives orders and does not fly, though he may have the brains of Solomon, will never get the results that a man will, who, day in and day out, leads his patrols over the line and infuses into his pilots the *espirit de corps*."
--Brigadier General William 'Billy' Mitchell, USAS, WWI

"Once committed to an attack, fly in at full speed."
 --Colonel Erich 'Bubi' Hartmann, German Air Force, WWII

"Maybe one out of ten good fighter pilots will be one of the hunters."
--Jack Ilfrey, USAAF, WWII

"There is only one seat in the cockpit of a fighter airplane."
--Richard Bach, *Stranger to the Ground*, 1963

"I mean, I had fast motor cars and fast motor bikes, and when I wasn't crashing airplanes, I was crashing motor bikes. It's all part of the game."
--Sir Harry Broadhurst, RAF, WWII

"We had a strict code of honor: you didn't shoot down a cripple and you kept it a fair fight."
--Captain Wilfrid Reid 'Wop' May, RFC, WWI

"Fight on and fly on to the last drop of blood and the last drop of fuel, to the last beat of the heart."
--Baron Manfred von Richthofen, German Air Force, WWI

"The first time I ever saw a jet, I shot it down."
--General Chuck Yeager, USAF, describing his first confrontation with a Me262 in WWII

"I look after the wingman. The wingman looks after me."
--Lt. Col. Francis S. "Gabby" Gabreski, USAF, WWII and Korea.

"The smallest amount of vanity is fatal in aeroplane fighting."
--Captain Edward V. 'Eddie' Rickenbacker, USAS, WWI

"Anybody who doesn't have fear is an idiot. It's just that you must make the fear work for you."
--Brigadier General Robin Olds, USAF, WWII and Vietnam

"Aerial gunnery is 90 percent instinct and 10 percent aim."
--Captain Frederick C. Libby, RFC, WWI

"Go in close, and when you think you are too close, go in closer."
--Major Thomas B. 'Tommy' McGuire, USAAF, WWII

"Speed is life."
--Captain Samuel Flynn, Jr., USN, WWII

"Aggressiveness was fundamental to success in air-to-air combat and if you ever caught a fighter pilot in a defensive mood you had him licked before you started shooting."
--Captain David McCampbell, USN, WWII

"Fly with the head and not with the muscles. The fighter pilot who is all muscle and no head will never live long enough for a pension."
--Colonel Willie Bats, German Air Force, WWII

"The air battle is not necessarily won at the time of the battle. The winner may have been determined by the amount of time, energy, thought and training an individual has previously accomplished in an effort to increase his ability as a fighter pilot."
--Colonel Gregory 'Pappy' Boyington, USMC, WWII

"A good fighter pilot, like a good boxer, should have a knockout punch"
--Captain Reade Tilley, USAAF, WWII

"He must have a love of hunting, a great desire to be the top dog."
--Sergei Dolgushin, Russian Air Force, WWII

"Know and use all the capabilities in your airplane. If you don't, sooner or later, some guy who does use them all will kick your ass."
--Lt Dave 'Preacher' Pace, USN Fighter Weapons School

"The first rule of all air combat is to see the opponent first."
--General Adolf Galland, German Air Force, WWII

"If you're in a fair fight, you didn't plan it properly."
--Nick Lappos, Chief R&D Pilot, Sikorsky Aircraft.

"Victory smiles upon those who anticipate the change in the character of war, not upon those who wait to adapt themselves after the changes occur."
--Giulio Douhet, *The Command of the Air.*

"No guts, no glory."
--General Frederick C. 'Boots' Blesse, USAF, Korea

"I think that the most important features of a fighter pilot are aggressiveness and professionalism. They are both needed to achieve the fighter pilot's goal."
--Colonel Gidi Livni, Israeli Air Force

"To be a good fighter pilot, there is one prime requisite — think fast, and act faster."
--Major John T. Godfrey, USAAF, WWII

Alternatives

You work in business sectors that seem to grow more challenging every year. Your company may have had great success in creating and delivering the very best technological solutions to value-conscious government customers. But with downward economic pressures it can be beneficial to shift focus from a singular "best" solution offering to provide customers with alternatives they can choose from. This might be in the shaping phase of capturing new business or the discussion phase of extending current business. It's true that you often know more about what the customer really needs than the customer does, but in saying so you risk coming across as arrogant. In many situations it's better to help customers explore as many alternatives and options as possible and patiently give them time to arrive at the best fit. And during these discussions you have a better chance at developing a more trust-based relationship, in which customers perceive you're acting in their best interest. This is actually the key to long sales.

During the sales process it's essential to understand how customers perceive themselves. It's conventional wisdom that government customers are now primarily interested in cost (price, actually). It's true that there are enormous federal budget problems and there's just not enough money to do all the things that need to be done. But it's also acknowledged that buying on the cheap usually means greater expenses in the future. Hamilton Associates did research on buyer "perceived value of differentiation" vs. "price sensitivity." The Convenience Buyer (low value perception and low price sensitivity) is not your sweet spot and the Loyal Buyer (high value perception and low price sensitivity) you've already convinced. It's the other types who should be more attractive to you.

Many government customers think of themselves as Price Buyers (low value perception and high price sensitivity). But it's difficult to be a Price Buyer when there aren't many competitors (perhaps that's some of the psychology behind Department of Defense initiatives attempting to increase competition in a monopsony market environment). Price Buyers tend to focus more on minimum requirements and price over price/feature tradeoffs. The challenge is to help Price Buyers understand all the many adjustments that could save them money in the long run. How? - By giving them alternatives so they can choose more wisely and become Value Buyers (high value perception and high price sensitivity). While price is important to Value Buyers, they carefully consider price/feature tradeoffs and are willing to pay more for higher value offerings that make sense. Satisfied Value Buyers have the potential to become your best loyal and advocate customers: those who buy repeatedly and refer you to others.

Alternatives also have benefit during the back end of the sales process. You've worked diligently with your customer to go through the details and now it's time for them to decide. After all this work you don't want to stand by and let them drift off. You need to help them help you close the deal. There are many time-tested closing approaches, but the use of alternatives in closing is simple and effective. Say, for example, the customer has a demonstrated need for a solution and they have funding. It's no longer a question of whether or not they will buy, but what they will buy and from whom. With an alternative closing approach you increase their sense of urgency and help them decide on one of your better solutions: "Would you want four or eight to begin with?" "Would you rather start with full training or the modified course?" "Which would better suit your needs: the vehicle with the communications suite or the basic tactical version?" It works.

Alternative approaches foster better customer discussions, guide the customer to the best fit solution, and help the customer make a good decision.

Alternative Quotes

"In war as in life, it is often necessary when some cherished scheme has failed, to take up the best alternative open, and if so, it is folly not to work for it with all your might."
--Winston Churchill

"History teaches us that men and nations behave wisely once they have exhausted all other alternatives."
--Abba Eban

"Old age is not so bad when you consider the alternatives."
--Maurice Chevalier

"If this nation is to be wise as well as strong, if we are to achieve our destiny, then we need more new ideas for more wise men reading more good books ... We must know all the facts and hear all the alternatives and

listen to all the criticisms."
--John Fitzgerald Kennedy

"When you have two alternatives, the first thing you have to do is to look for the third that you didn't think about, that doesn't exist."
--Shimon Peres

"In this world one is seldom reduced to make a selection between two alternatives. There are as many varieties of conduct and opinion as there are turns of feature between an aquiline nose and a flat one."
--Johann Wolfgang von Goethe

"Basically, everything I try to do is to present an alternative to what somebody else is doing."
--Matt Groening

"I never fit in. I am a true alternative. And I love being the outcast. That's my role in life, to be an outcast."
--Meat Loaf

"That's why I'm very proud of being American. I'm proud to pay taxes. I pay a lot of taxes, but it sure beats the alternative."
--Payne Stewart

"A long apprenticeship is the most logical way to success. The only alternative is overnight stardom, but I can't give you a formula for that."
--Chet Atkins

Authenticity

An essential element in gaining and maintaining good relations with customers is authenticity. But staying authentic can be difficult when you're under pressure to perform. Your behavior can change. You can hog the spotlight instead of sharing the glory with teammates. You can take over a customer meeting from better-informed colleagues. You can forget to do the right thing.

One way of reminding yourself to stay authentic is to listen to voices from the past. Not long ago my nephew sent me a 1943 Aircraft Worker's Manual from the Vega Aircraft Corporation, which he had found at a garage sale. This Burbank, California company was a Lockheed subsidiary during WWII and built famously sturdy Venturas, bomber patrol aircraft used by U.S. and British armed forces.

At the front of the book is an unattributed passage on *The Art of Getting Along*. The words on getting along were just as important to the leaders who wrote the manual as the specifications on how to properly rivet. The language is a little dated, but the thoughts are sincere and authentic.

THE ART OF GETTING ALONG
(Anonymous)
Vega Aircraft Corporation Worker's Manual, 1943

Sooner or later, a man, if he is wise, discovers that life is a mixture of good and bad, victory and defeat, give and take.

He learns that it doesn't pay to be a sensitive soul; that he should let some things go over his head like water off a duck's back.

He learns that he who loses his temper usually loses out and that all men have burnt toast for breakfast now and then, and that he shouldn't take the other fellow's grouch too seriously.

He learns that carrying a chip on his shoulder is the easiest way into a fight.

He learns that the quickest way to become unpopular is to carry tales and gossip about others and that buck passing always turns out to be a boomerang, and that it never pays.

He learns that every person is human and that it doesn't harm to smile and say, "Good Morning," even if it is raining.

He learns that most of the other fellows are as ambitious as he is, that they have intelligence that is as good or better, and that hard work and not cleverness is the secret of success.

He learns to sympathize with the youngster coming into business, because he remembers how bewildered he was when he first started out.

He learns not to worry when he loses an order, because experience has shown that if he always gives his best, his average will break pretty well.

He learns that folks are not any harder to get along with in one place than another and that the "getting along" depends about 98 percent on his own shoulder.

Authentic Quotes

"Hard times arouse an instinctive desire for authenticity."
--Coco Chanel

"To find yourself, think for yourself."
--Socrates

"Do what you can, with what you have, where you are."
--Theodore Roosevelt

"I've always tried to stay true to my authentic self. I think sometimes people project things on you, but I'm trying to handle everything that's happened to me with a certain amount of grace, dignity and good manners."
--Katie Couric

"This above all; to thine own self be true, and it must follow, as the night the day, thou canst not then be false to any man."
--William Shakespeare

"Be your authentic self. Your authentic self is who you are when you have no fear of judgment, or before the world starts pushing you around and telling you who you're supposed to be. Your fictional self is who you are when you have a social mask on to please everyone else. Give yourself permission to be your authentic self."
--Dr. Phil

"Be who you are and say what you feel because those who mind don't matter and those who matter don't mind."
--Dr. Seuss

"The successful, longtime partnerships almost always enjoy a true friendship, or at the very least an authentic regard for one another."
--Fred Couples

"Some writers confuse authenticity, which they ought always to aim at, with originality, which they should never bother about."
--W.H. Auden

"The weaker the data available upon which to base one's conclusion, the greater the precision which should be quoted in order to give the data

authenticity."
--Norman Ralph Augustine

"Always be a first-rate version of yourself instead of a second-rate version of someone else."
--Judy Garland

"Seek out that particular mental attribute which makes you feel most deeply and vitally alive, along with which comes the inner voice which says, 'This is the real me,' and when you have found that attitude, follow it."
--William James

"Don't compromise yourself. You are all you've got."
--Janis Joplin

"We are constantly invited to be who we are."
--Henry David Thoreau

"Music is your own experience, your thoughts, your wisdom. If you don't live it, it won't come out your horn."
--Charlie Parker

"Man will occasionally stumble over the truth, but usually manages to pick himself up, walk over or around it, and carry on."
--Winston Churchill

"Honesty and transparency make you vulnerable. Be honest and transparent anyway."
--Mother Theresa

"If you tell the truth you don't have to remember anything."
--Mark Twain

"The men who succeed best at public life are those who take the risk of standing by their own convictions."
--James A. Garfield

"There is no wisdom save in truth."
--Martin Luther

"Falsehoods not only disagree with truths, but usually quarrel among themselves."
--Daniel Webster

"You got to be who you are when you are."
--Snoop Dogg

"Example is not the main thing in influencing others. It is the only thing."
--Albert Schweitzer

"A man that seeks truth and loves it must be reckoned precious to any human society."
--Frederick the Great

"The truth - you've got to deal with it or it will kill you bit by bit."
--Ziggy Marley

"Best keep yourself clean and bright; you are the window through which you see the world."
--George Bernard Shaw

"If you do not tell the truth about yourself, you cannot tell it about other people."
--Virginia Woolf

"The truth is the kindest thing we can give folks in the end."
--Harriet Beecher Stowe

"Character is doing what's right when nobody is looking."
--J. C. Watts, Jr.

"If you seek authenticity for authenticity's sake you are no longer authentic."
--Jean-Paul Sartre

"To be what we are, and to become what we are capable of becoming, is the only end in life."
--Robert Louis Stevenson

"Your time is limited, so don't waste it living someone else's life."
--Steve Jobs

BASEBALL AND BUSINESS - BUYING

Baseball and Business

What is it about baseball that makes it so appealing, especially to those who appreciate strategy and finesse? A true fan once observed that he loved baseball because there was no clock and opportunities to win came equally and suddenly for both sides. The great game appeals to devotees of details and statistics, or baseball "arcana" (not "trivia," as George Will is quick to point out, for "there is nothing trivial in baseball.").

Business statistics are subject to similar scrutiny and admiration. Companies post sales and orders data, win/loss rates, profit and backlog and a host of other indicators to measure performance. Baseball comparisons can be drawn to business capture teams, where managers "suit up" like everyone else, where you "pitch" ideas to customers and "field" their questions, and where you have "home" (domestic) and "away" (international) games. And if you've worked in international business development like I have, you know that, just like baseball, it's a lot harder to win on the road than it is at home.

Did you see the 2011 movie *Moneyball*? It stars Brad Pitt as Oakland Athletics general manager Billy Beane, who devises a strategy to build a competitive team for the 2002 season despite Oakland's player payroll limitations. He enlists Peter Brand (played by Jonah Hill), a Yale economics whiz and baseball geek, to help him put statistical discipline into the recruiting and hiring process. Just as sales managers incur hostility from field representatives when a new numbers-focused system is introduced, Billy Beane gets vehement objections from his scouting staff when he hires undervalued players over players the scouts had "good feelings" about. Not to spoil it if you haven't seen the movie, *Moneyball* is a great story with many lessons for how to lead change in business.

At a Corporate Visions conference in Chicago in 2012, I had the chance to hear Billy Beane speak. He said his epiphany came when he realized that his own 1980 draft into the majors at number two was wrong because he was overvalued for things that didn't matter: what he was (tall, athletic, good-looking), rather than what he did (his performance statistics). In an interesting echo back to business development and sales, the Corporate Visions Power Messaging course teaches that your customer doesn't want to know as much about what your product *is* as much as what it *does*.

Beane's playing days played out quickly, but not his love for the game. Desperate to see his Athletics win, he and his protégé mined 150 years of baseball data, essentially mimicking Wall Street fund managers seeking an edge. Then they put their trust in the numbers and employed a business model to run the club like an insurance company. They found that the most important undervalued skill in a player was the ability to get on base. And despite the new Oakland lineup looking like "slow pitch softball players," they started winning games. Beane eliminated wasteful investment that did not contribute to winning and developed a strategy of "let the game come to you."

Moneyball is a brilliant little movie and the real Billy Beane is a great speaker. I came away from his presentation with many questions pointed directly at business statistics, such as: Are you using all the data available and are you measuring the right things? I believe companies sometimes track and measure things because they've always done it that way or simply because they can. Are there actionable data points in your customer relations satisfaction metrics and your sales and orders performance metrics? Are your metrics helping to drive your strategy?

In business, as in baseball, if you can get the numbers and players right you have a much better chance of winning. See you at the ballpark.

Baseball Quotes

"Baseball, it is said, is only a game. True. And the Grand Canyon is only a hole in Arizona."
--George F. Will

"Baseball is like a poker game. Nobody wants to quit when he's losing; nobody wants you to quit when you're ahead."
--Jackie Robinson

"Baseball is the only sport I know that when you're on offense, the other team controls the ball."
--Ken Harrelson

"Baseball is almost the only orderly thing in a very unorderly world. If you get three strikes, even the best lawyer in the world can't get you off."
--Bill Veeck

"Baseball is the only field of endeavor where a man can succeed three times out of ten and be considered a good performer."
--Ted Williams

"No matter how good you are, you're going to lose one-third of your games. No matter how bad you are you're going to win one-third of your games. It's the other third that makes the difference."
--Tommy Lasorda

"Show me a good loser in professional sports, and I'll show you an idiot."
--Leo Durocher

"A hot dog at the ballgame beats roast beef at the Ritz."
--Humphrey Bogart

"When you're in a slump, it's almost as if you look out at the field and it's one big glove."
--Vance Law

"When we played softball, I'd steal second base, feel guilty and go back."
--Woody Allen

"Good pitching will beat good hitting any time, and vice versa."
--Bob Veale, 1966

"It's hard to win a pennant, but it's harder losing one."
--Chuck Tanner

"During my 18 years I came to bat almost 10,000 times. I struck out about 1,700 times and walked maybe 1,800 times. You figure a ballplayer will

average about 500 at bats a season. That means I played seven years without ever hitting the ball."
--Mickey Mantle

"More than any other American sport, baseball creates the magnetic, addictive illusion that it can almost be understood."
--Thomas Boswell

"Why does everybody stand up and sing "Take Me Out to the Ballgame" when they're already there?"
--Larry Anderson

"I became a good pitcher when I stopped trying to make them miss the ball and started trying to make them hit it."
--Sandy Koufax

"Hitting is timing. Pitching is upsetting timing."
--Warren Spahn

"It ain't nothin' till I call it."
--Bill Klem, umpire

"Well, boys, it's a round ball and a round bat and you got to hit the ball square."
--Joe Schultz

"Pitchers, like poets, are born, not made."
--Cy Young

"You know you're pitching well when the batters look as bad as you do at the plate."
--Duke Snider

"Life will always throw you curves, just keep fouling them off... the right pitch will come, but when it does, be prepared to run the bases."
--Rick Maksian

"Been in this game one-hundred years, but I see new ways to lose 'em I never knew existed before."
--Casey Stengel

"After all my years, there are two things I've never got used to - haggling with a player over his contract and telling a boy he's got to go back."
--Connie Mack

"I never questioned the integrity of an umpire. Their eyesight, yes."
--Leo Durocher

"What counts aren't the number of double plays, but the ones you should have had and missed."
--Whitey Herzog

"Do what you love to do and give it your very best. Whether it's business or baseball, or the theater, or any field. If you don't love what you're doing and you can't give it your best, get out of it. Life is too short. You'll be an old man before you know it."
--Al Lopez

"When the ball is over the middle of the plate, the batter is hitting it with the sweet part of the bat. When it's inside, he's hitting it with the part of the bat from the handle to the trademark. When it's outside, he's hitting it with the end of the bat. You've got to keep the ball away from the sweet part of the bat. To do that, the pitcher has to move the hitter off the plate."
--Don Drysdale

"Stubbornness is usually considered a negative, but I think that trait has been a positive for me."
--Cal Ripken, Jr.

"Baseball is ninety percent mental. The other half is physical."
--Yogi Berra

"I'm a teammate guy, so whatever I can do to help my team to win like I have the past two years, that's what I want to do. If it takes for me to play first base, third base, right field, I just want to win the game."
--Albert Pujols

"I don't compare 'em, I just catch 'em."
--Willie Mays

"Awards mean a lot, but they don't say it all. The people in baseball mean more to me than statistics."
--Ernie Banks

"If I had my career to play over, one thing I'd do differently is swing more. Those 1,200 walks I got, nobody remembers them."
--Pee Wee Reese

"Why does everyone talk about the past? All that counts is tomorrow's game."
--Roberto Clemente

Buying

If you're in business development, you mostly think of selling. If you want to improve your win rate it's good to broaden your perspective and think about buying. A while ago I heard a compelling radio interview with Paco Underhill, a so-called "retail anthropologist," CEO of the Envirosell research firm, and author of "Why We Buy: The Science of Shopping." For many years he's studied the way people buy and believes at least some of the behavioral changes in consumption exhibited since the Great Recession may be here to stay. Here's a breakdown of his main points:

What's New
- **Overall consumption is down.** "Conspicuous consumers" are staying away from shopping malls as they save up, while others are taking newfound pride in how little they spend to buy or do things and how well they manage their existing budgets. The first group will return to their old ways once the economy picks up. The second group may not.
- **People are "buying down."** If they shopped at Nordstrom's, they now are shopping at Macy's. If they shopped at Macy's, they now are shopping at Target. If they shopped at Whole Foods, they now are shopping at Trader Joe's. If they were shopping at Trader Joe's, they now are shopping at Safeway. And so on.
- **There is greater buyer's remorse.** In a phenomenon called "deshopping," the closer people come to cash registers the more they're abandoning impulse selections. Some large stores are adding tables to hold the clutter (and hopefully get a follow-on buy from the next shopper in line).

- **There is greater impatience.** If an article in a stack of goods in a "big box" store is not marked, few shoppers will seek to purchase it. If there is a long line, they will walk out. Most big box stores will survive, but the ones that do will have to start providing better customer service.

- **A broad cross-section is affected.** These behavioral changes are occurring not only with people who were living beyond their means, but also with people who have money and are secure in their jobs. Everyone's shopping differently – from executives to manual laborers – as well as the newly employed and the retired.

What's Still the Same

- **The desire is still there.** Many people continue to shop, but "virtually." They go online and select things they really want to buy. But instead of ordering them now, they "bookmark" the pages for the future.

- **Brand names still work.** People are loyal to their favorite brands, but now look for ways to purchase them on sale or with coupons.

- **Payment schedules still matter.** If people decide on something they feel they really need (as well as want), an extended, low-interest payment schedule makes it easy for them to justify buying it now.

- **Value still counts.** There is a fair amount of hostility in recalling paying $150 for a pair of designer jeans and $27 for an exotic t-shirt in those flush times, especially if they did not last very long. Many people say they want prices to get "back to normal," but are willing to pay a little more if something "fits," won't go out of style, and will last a long time. This translates into "pride of ownership," much more fulfilling than the "thrill of acquisition."

- **Personal connections still apply.** In a sluggish economy, people tend to focus locally, not just to keep the money in the community, but to do business with people they know. People are now buying more food from farmer's markets. They're paying a little more but produce seems better being sold by those who grew it.

So what does this mean for the aerospace, defense, and government technology business? Think of these behaviors as they apply to your business relations. These "consumer" customers at home turn into your "administrative" customers when they enter their government workplaces. Although they're buying for someone else (warfighters, etc.), they can't help but be affected in their acquisition decisions by the lingering effects of the economic downturn and likely are making more conservative choices. Have you seen examples of this? Every time I hear sound bites from senior White House and DoD officials about the fate of programs, I hear echoes of recession thinking.

So what do you do? Being patient and understanding is a great starting point in helping your financially strapped government customers. Focus on what can be done now, while keeping options open for the future. Look for ways to stretch their budget dollars. Keep the options reasonable. Emphasize affordability, but also best value for their investments. Seek flexibility in payment schedules. Improve customer service. Maintain close customer relations: "We were with you then, we're with you now, and we'll be with you when we dig out of this economic mess." Instead of focusing on selling, you could shift to "helping our customers buy." It's subtle, but maybe it's just the right tone for recession-changed buyers.

Buying Quotes

"A budget tells us what we can't afford, but it doesn't keep us from buying it."
--William Feather

"A book worth reading is worth buying."
--John Ruskin

"Ask your child what he wants for dinner only if he's buying."
--Fran Lebowitz

"But paying is part of the game of life: it is the joy of buying that we crave."
--Gilbert Parker

"Buying is a profound pleasure."
--Simone de Beauvoir

"Certainly there are things in life that money can't buy, but it's very funny - Did you ever try buying them without money?"
--Ogden Nash

"Credit buying is much like being drunk. The buzz happens immediately and gives you a lift... The hangover comes the day after."
--Joyce Brothers

"No illusion is more crucial than the illusion that great success and huge money buy you immunity from the common ills of mankind, such as cars that won't start."
-- Larry McMurtry

"I do a lot of curiosity buying; I buy it if I like the album cover, I buy it if I like the name of the band, anything that sparks my imagination."
--Bruce Springsteen

"Whoever said money can't buy happiness simply didn't know where to go shopping."
--Bo Derek

"I went to a general store, but they wouldn't let me buy anything specific."
--Stephen Wright

"Lives, like money, are spent. What are you buying with yours?"
--Roy H. Williams

"Many a man thinks he is buying pleasure, when he is really selling himself to it."
--Benjamin Franklin

"Many an optimist has become rich by buying out a pessimist."
--Robert G. Allen

"Marrying a man is like buying something you've been admiring for a long time in a shop window. You may love it when you get it home, but it doesn't always go with everything else in the house."
--Jean Kerr

"There are all sorts of cute puppy dogs, but it doesn't stop people from going out and buying Dobermans."
--Angus Young

"Buy land. They're not making it anymore."
-- Mark Twain

"If you make a habit of buying things you do not need, you will soon be selling things you do."
--Filipino Proverb

"If you are buying a cow, make sure that the price of the tail is included."
--Tamil Proverb

"An inch of time is an inch of gold, but you cannot buy that inch time with an inch of gold."
--Chinese proverb

"A study of economics usually reveals that the best time to buy anything is last year."
--Marty Allen

"If stock market experts were so expert, they would be buying stock, not selling advice."
--Norman Ralph Augustine, aerospace business leader

"Only buy something that you'd be perfectly happy to hold if the market shut down for 10 years."
--Warren Buffett

"When buying shares, ask yourself, would you buy the whole company?"
--Rene Rivkin

"Business is not financial science, it's about trading … buying and selling. It's about creating a product or service so good that people will pay for it."
--Anita Roddick

"When somebody buys a stock it's because they think it's going to go up and the person who sold it to them thinks it's going to go down. Somebody's wrong."
--George Ross

"The nature of any human being, certainly anyone on Wall Street, is 'the better deal you give the customer, the worse deal it is for you'."
--Bernard Madoff

"Make your product easier to buy than your competition, or you will find your customers buying from them, not you."
--Mark Cuban

CHALLENGE YOUR CUSTOMER - COACHES - CONVERSATION - CREATE THE CUSTOMER - CUSTOMER-CENTRIC CULTURE

Challenge Your Customer

No, this doesn't mean challenging your customer to something like a duel. It means rethinking business development approaches in a down economy and trying proven sales behaviors that can lead to more new business. At a recent social gathering I had a long talk with the son of a close friend. Rick, whom I've watched grow up into an intelligent and articulate man, works for a California technology company. He told me about a book which his business development group profited from after only a few months of adopting its principles and techniques. I've read a lot of sales books and am skeptical of any such claims. But because I trust Rick, I read the book and found it compelling.

The Challenger Sale: Taking Control of the Customer Conversation was written by Matthew Dixon and Brent Adamson of the Sales Executive Council, a program within the Corporate Executive Board. It's based on research they compiled from interviewing and testing 6,000 sales professionals in all business sectors throughout the world. Their findings are praised by Neil Rackham, author of *Spin Selling*, the well-known reference book on complex sales.

Here are some background details from *The Challenger Sale*. Over the past decade buying patterns of large-account customers have changed. Instead of products, large-account administrative customers want solutions. And these big sales have become more complex, demanding broad-based consensus among stakeholders, increasing risk aversion on the part of key decision makers, and calling for greater customization.

This was all difficult enough, but then the economy went soft and sales started disappearing. So in their research, the authors also were searching for what kinds of behaviors salespeople needed in a down economy. They compiled all their data and found that observed sales behaviors fell into five profile models: The Challenger, The Relationship Builder, The Hard Worker, The Lone Wolf, and The Problem Solver. The authors assumed that The Relationship Builder profile would be best in a recession. What they discovered was that the Challenger profile was not only the best in a down economy, it was the clear winner for gaining new business in any environment - period. This behavioral profile dominates high performer salespeople, with Challengers representing more than 50 percent in the "star" category of all sales forces.

What distinguishes Challenger sales representatives from the others? The Challenger is strongest in these six of 44 tested attributes:

- Offers the customer unique perspectives
- Has strong two-way communication skills
- Knows the individual customer's value drivers
- Can identify economic drivers of the customer's business
- Is comfortable discussing money
- Can pressure the customer

The research showed that while the Relationship Builder is good for scouting opportunities and keeping current programs sold, the Challenger is best at getting new business and extending current business. The Relationship Builder focuses on resolving tension to make situations more amicable and encouraging collaboration. In contrast, the Challenger focuses on creating constructive tension to push the customers out of their comfort zone (the Call to Action). This is hard for many business development leaders to accept, for their inclination is to smooth over tension and keep the customer "happy." But a happy customer may not stay that way if needs aren't met. And needs can't be met if the customer can't be helped to make a decision to buy. The Challenger may not be liked by bosses as well as the Hard Worker or the Problem Solver and the Challenger may be as difficult to control as the Lone Wolf. Yet, the Challenger does something the others can't: win more business.

The Challenger does three things in relating to the customer: Teach, Tailor, and Take Control. In teaching the customer, the Challenger is an authority: not arrogant, but extremely knowledgeable about products, services, markets, and competitors. The Challenger engages in robust two-way conversations, challenges customer assumptions, and gives insight to reframe customer perspectives. In tailoring approaches to the customer set, the Challenger delivers the right message to the right person, generates widespread support, and is easy to work with. It's in the Challenger mode of taking control of the sale that is hardest for many business developers to emulate. But if you don't take positive control of the sale, your competitors will.

To a certain extent Challenger behaviors can be taught and learned. Challengers can be made, not just born. But if it doesn't feel right, don't do it. There is no benefit and pleasure from trying to be something you're not and there are plenty of other positions where you can contribute to business development and grow your company. But if you have an inner Challenger and your customer doesn't have defined requirements and your boss and organization are supportive, try and step up your game. Remember: It's not what you sell, it's how you sell it.

Challenge Quotes

"The ultimate measure of a man is not where he stands in moments of comfort and convenience, but where he stands at times of challenge and controversy."
--Martin Luther King, Jr.

"We must accept life for what it actually is - a challenge to our quality without which we should never know of what stuff we are made, or grow to our full stature."
--Robert Louis Stevenson

"Don't be afraid to challenge the pros, even in their own backyard."
--Colin Powell

"How you respond to the challenge in the second half will determine what you become after the game, whether you are a winner or a loser."
--Lou Holtz

"The challenge is in the moment; the time is always now."
--James A. Baldwin

"The only use of an obstacle is to be overcome. All that an obstacle does with brave men is, not to frighten them, but to challenge them."
--Woodrow Wilson

"My dad always used to tell me that if they challenge you to an after-school fight, tell them you won't wait-you can kick their ass right now."
--Cameron Diaz

"I'm hungrier than those other guys out there. Every rebound is a personal challenge."
--Dennis Rodman

"I thoroughly disapprove of duels. If a man should challenge me, I would take him kindly and forgivingly by the hand and lead him to a quiet place and kill him."
--Mark Twain

"Accept the challenges so that you may feel the exhilaration of victory."
--General George S. Patton

"The greatest challenge to any thinker is stating the problem in a way that will allow a solution."
--Bertrand Russell

"When you've got something to prove, there's nothing greater than a challenge."
--Terry Bradshaw

"There are no great people in this world, only great challenges which ordinary people rise to meet."
--Admiral William Frederick Halsey, Jr.

"Never do things others can do and will do if there are things others cannot do or will not do."
--Amelia Earhart

"It isn't the mountains ahead to climb that wear you out, it's the pebble in your shoe."
--Muhammad Ali

"I've learned that you can tell a lot about a person by the way he/she handles these three things: a rainy day, lost luggage, and tangled Christmas tree lights."
--Maya Angelou

"To truly laugh, you must be able to take your pain, and play with it!"
--Charlie Chaplin

"The pessimist sees difficulty in every opportunity. The optimist sees the opportunity in every difficulty."
--Winston Churchill

"When I am working on a problem, I never think about beauty. I think only how to solve the problem. But when I have finished, if the solution is not beautiful, I know it is wrong."
--Buckminster Fuller

"Great minds have purposes, little minds have wishes. Little minds are subdued by misfortunes, great minds rise above them."
--Washington Irving

"When you first start off trying to solve a problem, the first solutions you come up with are very complex, and most people stop there. But if you keep going, and live with the problem and peel more layers of the onion off, you can often times arrive at some very elegant and simple solutions."
--Steve Jobs

"Obstacles don't have to stop you. If you run into a wall, don't turn around and give up. Figure out how to climb it, go through it, or work around it."
--Michael Jordan

"The most rewarding things you do in life are often the ones that look like they cannot be done."
--Arnold Palmer

"If you want the rainbow, you've got to put up with the rain."
--Dolly Parton

"Every problem has in it the seeds of its own solution. If you don't have any problems, you don't get any seeds."
--Norman Vincent Peale

"All misfortune is but a stepping stone to fortune."
--Henry David Thoreau

"What seems to us bitter trials are often blessings in disguise."
--Oscar Wilde

"Other people and things can stop you temporarily. You're the only one who can do it permanently."
--Zig Ziglar

"You may not realize it when it happens, but a kick in the teeth may be the best thing in the world for you."
--Walt Disney

"Life is a challenge, meet it."
--Mother Theresa

Coaches

Do you have a coach to help you win business? I don't mean a manager who coaches you on your sales techniques or a mentor who encourages your efforts, but someone inside or near your customer set. Someone who is not a consultant, but consults with you. Someone who wants you to win in a straightforward and honest way that will solve problems and get an organization what it needs to fulfill its mission. It doesn't matter if you're shaping, capturing, or extending business; a coach can be critical to your success. The difference between a business coach and a sports coach is that in business your coach is on the field, not on the sidelines.

According to Miller Heiman, a good coach is:
- Credible within the customer's organization
- Knowledgeable of the organization's requirements
- A person with whom you have credibility
- Someone who wants you to get the job
- Inside or close to the customer's organization

A Harvard Business School study shows that, on average, it takes approximately 12 contacts with a decision maker to close a substantive deal. Seven of those interactions must be "quality" contacts: face-to-face discussions, lengthy phone conversations, or active electronic messaging. Only 10 percent of business developers make more than five contacts before they move on. Many times they can't even get in the door to talk with the right executives and officials. A great way to shortcut the 12-contact rule is to use a coach. Without a coach you only have a 20 percent chance of getting that first critical meeting. With a coach, you have an 80 percent chance of a key decision maker agreeing to talk with you.

Finding a coach (adapted from *Codebreakers*):

- Use your current customer base to help with new prospects
- Use your vendors, suppliers and consultants
- Ask referral sources to be coaches
- Ask people in your own organization for suggestions
- Join forces with a partner company business developer

This aerospace, defense and technology business environment is tough. Get a competitive edge by getting a coach.

Coach Quotes

"People of mediocre ability sometimes achieve outstanding success because they don't know when to quit."
--George Allen

"The only correct actions are those that demand no explanation and no apology."
--Red Auerbach

"You can't think and hit at the same time."
--Yogi Berra

"Don't go to the grave with life unused."
-- Bobby Bowden

"Never quit. It is the easiest cop-out in the world. Set a goal and don't quit until you attain it. When you do attain it, set another goal, and don't quit until you reach it. Never quit."
--Bear Bryant

"A tie is like kissing your sister."
--Duffy Daugherty

"No one comes into our house and pushes us around."
--Dan Devine

"Before you can win, you have to believe you are worthy."
--Mike Ditka

"If I were playing third base and my mother were rounding third with the run that was going to beat us, I'd trip her. Oh, I'd pick her up and brush her off and say, 'Sorry, Mom,' but nobody beats me."
--Leo Durocher

"We'll take what the other team gives us. We'll scratch where it itches."
--Hayden Fry

"People who enjoy what they are doing invariably do it well."
--Joe Gibbs

"Find out what the other team wants to do. Then take it away from them."
--George Halas

"When in doubt, punt!"
--John Heisman

"No one has ever drowned in sweat."
--Lou Holtz

"No one plays this or any game perfectly. It's the guy who recovers from his mistakes who wins."
--Phil Jackson

"The difference between ordinary and extraordinary is that little extra."
--Jimmy Johnson

"Mental toughness is to physical as four is to one."
--Bobby Knight

"Pressure is what you feel when you don't know what's going on."
--Chuck Knoll

"The truth is that many people set rules to keep from making decisions."
--Mike Krzyzewski

"Setting a goal is not the main thing. It is deciding how you will go about achieving it and staying with that plan."
--Tom Landry

"The difference between the possible and the impossible lies in a person's determination."
--Tommy Lasorda

"The measure of who we are is what we do with what we have."
--Vince Lombardi

"The only yardstick for success our society has is being a champion. No one remembers anything else."
--John Madden

"All coaching is, is taking a player where he can't take himself."
--Bill McCartney

"You can never work too hard on attitudes, effort and technique."
--Don Meyer

"The only discipline that lasts is self-discipline."
--Bum Philips

"Look for your choices, pick the best one, then go with it."
--Pat Riley

"One loss is good for the soul, too many losses is not good for the coach."
--Knute Rockne

"You never lose a game if the opponent doesn't score."
--Darrell Royal

"The one thing that I know is that you win with good people."
--Don Shula

"If you make every game a life and death proposition, you're going to have problems. For one thing, you'll be dead a lot."
--Dean Smith

"Most games are lost, not won."
--Casey Stengel

"Yesterday is a cancelled check. Today is cash on the line. Tomorrow is a promissory note."
--Hank Stram

"The key step for an infielder is the first one, to the left or right, but before the ball is hit."
--Earl Weaver

"You are only as good as the coach thinks you are."
--Brian Williams

"It's the little details that are vital. Little things make big things happen."
--John Wooden

Conversation

I was thumbing through the *Economist* on a long flight and came across an article about the art of conversation. With all the emphasis placed on listening to the customer, it's easy to overlook the importance of what you say and how you say it when it's your turn to speak. The article ("Chattering Classes") went into great historical and cultural detail on the subject, but what struck me most was how contemporary the advice from Cicero sounded. For example, in his 44 B.C. essay *On Duties*, he pointed out that it was rude to interrupt another speaker, as good conversation comes from the participants alternating meaningful comments. Cicero went on to say that since no one had yet set down rules for conversation he would do so and created a list that wouldn't be out of place on any self-help website.

The rules we learn from Cicero are these:
1. Speak clearly
2. Speak easily, but not too much, especially when others want their turn
3. Do not interrupt
4. Be courteous
5. Deal seriously with serious matters and gracefully with lighter ones
6. Never criticize people behind their backs
7. Stick to subjects of general interest
8. Do not talk about yourself
9. And, above all, never lose your temper

Rules two and three drew my interest regarding interactions with customers. Great conversations have flow, with each participant taking part and contributing something of interest. Have you ever been in a customer meeting and one of your extrovert colleagues kept talking and talking and talking and then, when the customer interjected a comment, your colleague interrupted him and continued on? Yikes! Lesson number one for every business developer should be to never, ever interrupt the customer.

Cicero's rule six is also very important to remember. It's not just that what goes around comes around, it's that the customer, whether consciously or subconsciously, will suspect you may say something about him or her in another setting. That perception will silence the customer and reduce any trust assurances you've built up. However, if your customer starts criticizing your competition, that's another matter altogether. Refer back to rule three

Rule eight was intended for general conversation, but you might want to disclose to your customer certain of your own experiences and qualifications which show you know what you're talking about. For the customer first "buys" you, then your company and finally your solution. If they don't know you and trust your company, there's little chance they'll buy your solution, no matter how brilliant it is.

So the next time you've thoroughly listened to your customer and are about speak, think first of Cicero and you might make an impression for the ages.

Conversation Quotes

"Conversation should be pleasant without scurrility, witty without affectation, free without indecency, learned without conceitedness, novel without falsehood."
--William Shakespeare

"A man's character may be learned from the adjectives which he habitually uses in conversation."
--Mark Twain

"The hardest thing about being famous is that people are always nice to you. You're in a conversation and everybody's agreeing with what you're saying - even if you say something totally crazy. You need people who can tell you what you don't want to hear."
--Al Pacino

"A conversation is a dialogue, not a monologue. That's why there are so few good conversations: due to scarcity, two intelligent talkers seldom meet."
--Truman Capote

"My idea of good company is the company of clever, well-informed people who have a great deal of conversation; that is what I call good company."
--Jane Austen

"If you're yelling you're the one who's lost control of the conversation."
--Taylor Swift

"People say conversation is a lost art; how often I have wished it were."
--Edward R. Murrow

"Conversation about the weather is the last refuge of the unimaginative."
--Oscar Wilde

"Silence is one of the great arts of conversation."
--Marcus Tullius Cicero

"If other people are going to talk, conversation becomes impossible."
--James Whistler

"There is no conversation more boring than the one where everybody agrees."
--Michel de Montaigne

"There are those moments when you shake someone's hand, have a conversation with someone, and suddenly you're all bound together because you share your humanity in one simple moment."
--Ralph Fiennes

"The reason why so few people are agreeable in conversation is that each is thinking more about what he intends to say than others are saying."
--Francois de La Rochefoucauld

"She had lost the art of conversation but not, unfortunately, the power of speech."
--George Bernard Shaw

"The real art of conversation is not only to say the right thing at the right place but to leave unsaid the wrong thing at the tempting moment."
--Dorothy Nevill

"One of the best rules in conversation is, never to say a thing which any of the company can reasonably wish had been left unsaid."
--Jonathan Swift

"Well I do think, when there are more women, that the tone of the conversation changes, and also the goals of the conversation change. But it doesn't mean that the whole world would be a lot better if it were totally run by women. If you think that, you've forgotten high school."
--Madeleine Albright

"Women speak because they wish to speak, whereas a man speaks only when driven to speak by something outside himself like, for instance, he can't find any clean socks."
-- Jean Kerr

"It was impossible to get a conversation going, everybody was talking too much."
--Yogi Berra

"The great secret of succeeding in conversation is to admire little, to hear much; always to distrust our own reason, and sometimes that of our friends; never to pretend to wit, but to make that of others appear as much as possibly we can; to hearken."
--Benjamin Franklin

"Wit is the salt of conversation, not the food."
--William Hazlitt

"Conversation between Adam and Eve must have been difficult at times because they had nobody to talk about."
--Agnes Repplier

"In conversation, humor is worth more than wit and easiness more than knowledge."
--George Herbert

"No one appreciates the very special genius of your conversation as the dog does."
--Christopher Morley

"I was diagnosed a number of years ago with obsessive-compulsive disorder - which everyone has, to some degree - and I have this really annoying trait where in conversations, I always steer it back to something that happened to me."
--Paula Poundstone

"It is all right to hold a conversation, but you should let go of it now and then."
--Richard Armour

"I feel like my first conversation with someone, I really get a good feeling about who that person is and mainly about how open they are."
--Elisabeth Shue

"Inject a few raisins of conversation into the tasteless dough of existence."
--O. Henry

Create the Customer

For decades Peter Drucker was a leading guru of business management and organizational theory. Perhaps you remember some of his classic observations. My favorite is: "There is only one valid definition of business purpose: to create a customer." It is so elegantly truthful.

Drucker's *The Practice of Management* was voted one of the most influential management books of the 20th Century and his principles hold true. He challenged those who wished to grow their businesses to first ask themselves these basic questions, the answers to which seem simple, but aren't: (1) What is our mission? (2) Who is our customer? and (3) What does our customer value?

What is our mission?

Fundamentally, your business should be what you do well, what your core competencies are. That doesn't mean you should restrict yourself only to current types of programs, but it does acknowledge that if you want to move into adjacent markets you need to adjust your people and processes. Companies confuse their identities all the time. Remember years ago when Northrop Grumman set out to be a large systems integrator, but wound up owning the largest shipbuilding capacity in the Western Hemisphere? Remember the indecision at Hewlett-Packard over whether or not they would be in the computer business? Remember when Starbucks forgot it was all about the coffee? Many aerospace, defense, and technology companies have 100 years of heritage. What should your line of business look like in the second 100 years?

Who is our customer?

You don't produce and sell things for the consumer-oriented public. You have sets of administrative customers who buy for others. Your programs can have at least six types of customers – operators, maintainers, managers, owners, funders, and deciders – all with differing visions of what they want. To stay in business you need to satisfy them all and that can be difficult. Most of your customers work for governments, both here in the U.S. and abroad. Where you do have commercial customers, they are concentrated mostly in heavily regulated industries, such as commercial aviation, telecommunications, and IT services. Do you know your customers now? Do you know who your future customers could be?

What does our customer value?

The previous two questions require introspection and analysis and they can be hard to answer, but they're not as hard as this one. To answer this question properly you have to really understand your customers. The most difficult concept to comprehend is that if the customer doesn't value your product or service it doesn't matter how much time and resources you've invested in it. Supply without demand is a dead end. That means you should pay attention to the demand side. I recall a former colleague ending each e-mail with the tag line "Create the Need." I thought it a bit pushy when I first saw it, then began to realize that he was on to something. Many customers don't know what they would consider valuable because they haven't been afforded insight into possibilities. As Henry Ford famously said, "If I had asked people what they wanted, they would have said faster horses." Recall the launch of Steve Jobs' iPad. Most of the initial reaction was negative because people had to start using it to fully realize how great it was. Ford and Jobs knew what customers *would* value, not what they valued at the moment. What's on deck in your line of business that customers would value?

If you want to grow business, you need to create customers.

Create Quotes

"The best way to predict the future is to create it."
--Peter Drucker

"Trust yourself. Create the kind of self that you will be happy to live with all your life."
--Golda Meir

"To hell with circumstances; I create opportunities."
--Bruce Lee

"The human body has two ends on it: one to create with and one to sit on. Sometimes people get their ends reversed."
--Theodore Roosevelt

"Imagination is the beginning of creation. You imagine what you desire, you will what you imagine and at last you create what you will."
--George Bernard Shaw

"There is nothing like a dream to create the future."
--Victor Hugo

"To create something you must be something."
--Johann Wolfgang von Goethe

"Engineers like to solve problems. If there are no problems handily available, they will create their own problems."
--Scott Adams

"Those who create are rare."
--Coco Chanel

"If knowledge can create problems, it is not through ignorance that we can solve them."
--Isaac Asimov

"I think that God, in creating man, somewhat overestimated his ability."
--Oscar Wilde

"Desperation is a necessary ingredient to learning anything, or creating anything. Period. If you ain't desperate at some point, you ain't interesting."
--Jim Carrey

"You cannot create experience. You must undergo it."
--Albert Camus

"The significant problems we face cannot be solved by the same level of thinking that created them."
--Albert Einstein

"*Begin with the end in mind* is based on the principle that all things are created twice. There's a mental or first creation, and a physical or second creation to all things."
--Stephen R. Covey

"If you want to make an apple pie from scratch, you must first create the universe."
--Carl Sagan

"Creativity is just connecting things. When you ask creative people how they did something, they feel a little guilty because they didn't really do it, they just saw something. It seemed obvious to them after a while. That's because they were able to connect experiences they've had and synthesize new things."
--Steve Jobs

"Don't think. Thinking is the enemy of creativity. It's self-conscious, and anything self-conscious is lousy. You can't try to do things. You simply must do things."
--Ray Bradbury

"Anxiety is the hand maiden of creativity."
--T. S. Eliot

"Creativity comes from a conflict of ideas."
--Donatella Versace

"The chief enemy of creativity is 'good' sense."
--Pablo Picasso

"An essential aspect of creativity is not being afraid to fail."
--Edwin Land

"Creativity is more than just being different. Anybody can plan weird; that's easy. What's hard is to be as simple as Bach. Making the simple, awesomely simple, that's creativity."
--Charles Mingus

"If you go with your instincts and keep your humor, creativity follows. With luck, success comes, too."
--Jimmy Buffett

"I go wherever my creativity takes me."
--Lil Wayne

"A hunch is creativity trying to tell you something."
--Frank Capra

"Trying to force creativity is never good."
--Sarah McLachlan

"Creativity takes courage."
--Henri Matisse

"Creativity is piercing the mundane to find the marvelous."
--Bill Moyers

"You can't use up creativity. The more you use, the more you have."
--Maya Angelou

"Creativity requires the courage to let go of certainties."
--Erich Fromm

"The thing is to become a master and in your old age to acquire the courage to do what children did when they knew nothing."
--Ernest Hemingway

"Creativity is an act of defiance."
--Twyla Tharp

"Despite a lack of natural ability, I did have the one element necessary to all early creativity: naïveté, that fabulous quality that keeps you from knowing

just how unsuited you are for what you are about to do."
--Steve Martin

Customer-Centric Culture

Does your company have a customer-centric culture? Where the customer comes first in everything you do? Not many companies do, especially aerospace, defense, and technology companies selling to government customers. There are lots of reasons for this. It can seem unnatural, even artificial, in a business involving complex programs and millions of dollars.

Many business leaders falsely believe that if you create brilliant solutions, perform well on contracts, and stick to the fundamentals of wise cash deployment, all customer issues will work out by themselves. When problems don't work out well on their own, leaders reexamine program progress, mission success, win rates, CPAR ratings, and other metrics. The interesting thing about all this is your company may be performing well, but if the customer doesn't *feel* you're performing well, then you aren't.

I believe the key to building a customer-centric culture in any company is not in the numbers, but in the leadership. CEOs and presidents need to foster an atmosphere that encourages everyone in the company to believe in the importance and value of what they are doing and hold their leadership accountable for delivering excellence in customer relations. Everyone in the company is capable of understanding and getting behind this culture shift if it's clearly stated, led from the top and spread throughout the enterprise without exception.

In customer relations training sessions with engineers, I've been surprised at how they embrace this concept, once they're shown evidence that in business, feelings are facts. If you improve the relationship you have with your customer, improvement in your program ratings will follow. I asked class participants to come up with attributes they think customers would value in a company with a customer-centric culture. Here's a composite list from multiple sessions in order of frequency:

Top Twenty Customer-Centric Cultural Attributes
- Trustworthy
- Responsive
- Attentive

- Partnering
- Honest
- Team-Oriented
- Visionary
- Committed
- Respectful
- Focused
- Excellent
- Flexible
- Responsible
- Cooperative
- Effective
- Friendly
- Affordable
- Dependable
- Agile
- Enjoyable

And now for some customer-centric quotes from business leaders and sales experts.

Customer-Centric Culture Quotes

"He profits most who serves best."
--Arthur F. Sheldon

"America is ripe for a service revolution."
--Harvey Mackay

"Ask your customers to be part of the solution, and don't view them as part of the problem."
--Alan Weiss

"Being on par in terms of price and quality only gets you into the game. Service wins the game."
--Tony Alessandra

"Biggest question: Isn't it really 'customer helping' rather than customer service? And wouldn't you deliver better service if you thought of it that

way?"
--Jeffrey Gitomer

"Business is not just doing deals; business is having great products, doing great engineering, and providing tremendous service to customers."
--Ross Perot

"Consumers are statistics. Customers are people."
--Stanley Marcus

"Customers long to interact with - even relate to - employees who act like there is still a light on inside."
--Chip Bell

"Customers today want the very most and the very best for the very least amount of money, and on the best terms. Only the individuals and companies that provide absolutely excellent products and services at absolutely excellent prices will survive."
--Brian Tracy

"Don't try to tell the customer what he wants. If you want to be smart, be smart in the shower. Then get out, go to work and serve the customer!"
--Gene Buckley

"Every great business is built on friendship."
--J.C. Penney

"Forget about the sales you hope to make and concentrate on the service you want to render."
--Harry Bullis

"Here is a simple but powerful rule: always give people more than what they expect to get."
--Nelson Boswell

"If the shopper feels like it was poor service, then it was poor service. We are in the customer perception business."
--Mark Perrault

"If you do build a great experience, customers tell each other about that. Word of mouth is very powerful."
--Jeff Bezos

"If you don't care, your customer never will."
--Marlene Blaszczyk

"If you don't genuinely like your customers, chances are they won't buy."
--Tom Watson

"If you get everybody in the company involved in customer service, not only are they 'feeling the customer' but they're also getting a feeling for what's not working."
--Penny Handscomb

"If you're not serving the customer, your job is to be serving someone who is."
--Jan Carlzon

"It is not the employer who pays the wages. Employers only handle the money. It is the customer who pays the wages."
--Henry Ford

"Make a customer, not a sale."
--Katherine Barchetti

"One customer well taken care of could be more valuable than $10,000 worth of advertising."
--Jim Rohn

"Our customers should take joy in our products and services."
--W. Edwards Deming

"People expect good service but few are willing to give it."
--Robert Gately

"Quality in a service or product is not what you put into it. It is what the client or customer gets out of it."
--Peter Drucker

"Revolve your world around the customer and more customers will revolve around you."
--Heather Williams

"The customer's perception is your reality."
--Kate Zabriskie

"The goal as a company is to have customer service that is not just the best but legendary."
--Sam Walton

"The longer you wait, the harder it is to produce outstanding customer service."
--William H. Davidow

"There are only two ways to get a new customer: 1. Solicit a new customer any way you can. 2. Take good care of your present customers, so they don't become someone else's new customers."
--Ed Zeitz

"You'll never have a product or price advantage again. They can be easily duplicated, but a strong customer service culture can't be copied."
--Jerry Fritz

"Would you do business with you?"
--Linda Silverman Goldzimer

DIFFERENTIATION IN COMPETITIONS - DAY IN THE LIFE OF A CUSTOMER

Differentiation in Competitions

You live in a competitive world and work in a competitive environment. For the past decade the aerospace, defense, and technology sectors have been doing pretty well. But now you're faced with declining acquisition budgets, looming deficits, and challenging customer attitudes. The already stiff competition in your industry has become fierce and it's quite possible that some companies may not survive. The ones that do will have to demonstrate and deliver value to customers and find ways to separate themselves from the pack.

All major aerospace, defense, and technology companies look pretty much the same to many government customers. Even the radio ads sound similar. Have you ever been in a capture review or Gold Team and looked at what was written for the value proposition? Did you notice that it didn't look any different from the competitor's value proposition? It didn't even look any different from what you would say about any other company product. If you don't know and can't articulate why you are different, how can the customer? Why would they pick you?

Research done a while back by Professor Noriaki Kano on product development and customer satisfaction showed the importance of putting some "Wow" in your offerings. When responding to a customer's request for proposal, you have to show the basic, or entry level, factors that allow your offer to be technically acceptable. These have no upside and can even be seen negatively if you can't prove they are fully compliant with what the customer has written into the RFP. You also have the opportunity to show performance factors (specifications, cost, schedule, etc.). These can have an upside if they are demonstrably superior to competitor offerings, but in many cases solutions offered by competitors are not all that different. In whatever way you can, you need to show the customer "delighters," specifications and attributes of your offering that can be perceived as valuable in comparison to others. In today's fierce markets, no matter if you are selling jet fighters or IT support, you have to be able to put greater customer-oriented value impact into your products and solutions. You need to show that your offerings have something more, whether it's ease of operation, greater ROI, that they are built with open-ended technical growth pathways - something that can separate you from the competition in the mind of the customer. And if what your offering does has greater value toward customer expectations – the "Wow" factor – then the customer will pick you.

The Corporate Visions people who run Power Messaging courses explain that most companies instinctively compete where there is value parity, where customer needs, your offerings, and competitor offerings overlap. That's a mistake. What happens in value parity? It becomes a cost shoot-out. Do you like being in cost shoot-outs? Does your company do well in cost shoot-outs? The key to success in winning is to position your offerings where they meet customer needs, but are clearly distinguishable from competitor offerings. You help do this by identifying three aspects to any selling point you make. Each point should be: (1) of value to the customer, (2) easy to defend, and (3) unique to you. If you are not selling something that is truly unique to you, then you're trapped in the gray muddle of basic and performance factors and you've reduced your chances of selection.

By the way, your toughest competitor in most cases isn't another company. It's indecision on the part of an information-overloaded customer who can't differentiate one solution from another. So in addition to being excellent, be different. If you're both, there's no competition.

There's probably no sport where it's more intensely competitive and harder to be different than basketball. Your team doesn't have to have a big sports budget so there's a low economic barrier to entry. You only have to have a few good players so there's no scale advantage. And you have to play a lot of games so lucky breaks don't count as much as, say, in football. Okay, this transition is a stretch, but anyway - here are some quotes to illustrate how basketball players and coaches differentiate themselves and their teams.

Basketball Quotes

"A basketball team is like the five fingers on your hand. If you can get them all together, you have a fist. That's how I want you to play."
--Mike Krzyzewski

"I can accept failure, everyone fails at something. But I can't accept not trying."
--Michael Jordan

"Basketball is like war in that offensive weapons are developed first, and it always takes a while for the defense to catch up."
--Red Auerbach

"If you're a basketball player, you've got to shoot."
--Oscar Robertson

"Left hand, right hand, it doesn't matter. I'm amphibious."
--Charles Shackleford

"It wasn't about the X's and the O's and the strategy; it was more about keeping 12 guys focused and committed to a task. That group dynamic, and then helping them to grow as people and basketball players."
--Isiah Thomas

"You can run a lot of plays when your X is twice as big as the other guy's O. It makes your X's and O's pretty good."
--Paul Westphal

"You will always miss 100% of the shots you don't take."
--Larry Byrd

"When you build bridges you can keep crossing them."
--Rick Pitino

"We have a great bunch of outside shooters. Unfortunately, all our games are played indoors."
--Weldon Drew

"You can't get much done in life if you only work on the days when you feel good."
--Jerry West

"When I dunk, I put something on it. I want the ball to hit the floor before I do."
--Darryl Dawkins

"Love never fails; Character never quits; and with patience and persistence; Dreams do come true."
--Pete Maravich

"It's what you get from games you lose that is extremely important."
--Pat Riley

"No matter what business you're in, you can't run in place or someone will pass you by."
--Jim Valvano

"You don't play against opponents. You play against the game of basketball."
--Bobby Knight

"Ego is the drug of stupidity."
--Tom Amberry

"We're shooting 100 percent - 60 percent from the field and 40 percent from the free-throw line."
--Norm Stewart

"They say that nobody is perfect. Then they tell you practice makes perfect. I wish they'd make up their minds."
--Wilt Chamberlain

"Don't let what you cannot do interfere with what you can do."
--John Wooden

"I'd rather have more heart than talent any day."
--Allen Iverson

"The secret is to have eight great players and four others who will cheer like crazy."
--Jerry Tarkanian

"Don't let what other people think decide who you are."
--Dennis Rodman

"There are really only two plays: *Romeo and Juliet*, and put the darn ball in the basket."
--Abe Lemons

"If you make every game a life and death proposition, you're going to have problems. For one thing, you'll be dead a lot."
--Dean Smith

"In basketball, the first person to touch the ball shoots it. Either that or the coach carefully diagrams a set play and then the first player to touch it shoots it."
--Gene Klein

"The game is too long, the season is too long and the players are too long."
--Jack Dolph

"We can't win at home. We can't win on the road. As general manager, I just can't figure out where else to play."
--Pat Williams

"Basketball is like photography, if you don't focus, all you have is the negative."
--Dan Frisby

"One day of practice is like one day of clean living. It doesn't do you any good."
--Abe Lemmons

"Never mistake activity for achievement."
--Bill Walton

"If you meet the Buddha in the lane, feed him the ball."
--Phil Jackson

"The idea is not to block every shot. The idea is to make your opponent believe that you might block every shot."
--Bill Russell

"The only difference between a good shot and a bad shot is if it goes in or not."
--Charles Barkley

Day in the Life of a Customer

A "Day in the Life of a Customer" exercise can help you conduct more effective value analysis and create better customer messaging. This informal activity can best be done by people on your team who have been customers or know a similar customer so well that they can "channel" the persona. I recall a case study about an agricultural chemical company that was losing market share in a very competitive market space. They were seeking better ways to move their chemicals through independent distributors. But there was very little differentiation among supplier products and the market had devolved into a complex, rebate-rich network that no one was happy with: suppliers, distributors or farmers.

One of the company leaders had been a farmer, a third-generation farmer, in fact. At an offsite, he began a monologue about how hard it was to be a farmer, how much he had to know, how dependent he was on the weather, how his crops were in constant danger from pests, and how frustrating it was to have to orchestrate all the support needed on a modern, large farm. The rest of the company leaders were following the story intently and then it hit them like a summer lightning bolt. The better business was in chemical services, not chemical supplies. They began to offer customized chemical services at affordable prices through their distributor partners and their business took off.

Let's extend this "Day in the Life of a Customer" concept to the federal market. Jane is a Navy captain and head of a large group responsible for surface warfare program support. She's a well-respected senior officer eligible for flag rank and has had a good and challenging career. At the Naval Academy she first learned that it wasn't enough to be smart and quick; she had to prove herself daily to get her opinions heard. Through determination and hard work Jane advanced quickly, earned trust, and was given ever-greater responsibilities. She married a good man and together they found ways to juggle two careers and two daughters. But the girls were older and more complicated and her husband was getting restless, agitated at having to keep putting off going back home to take over his ailing father's business. Jane's parents weren't doing too well either. Her father was recently diagnosed with dementia and her mother was worried that she wouldn't be able to care for him. These problems were never far away.

Jane rose early one morning and went for a jog around the neighborhood. Her morning run was the only time that was hers and she let thoughts come in and out as she ran under the glowing street lamps. She loved the Navy and all the good and bad that came with it. The Kingfisher Program, however, was giving her fits. It was a special program funded quickly to meet an urgent operational need and it was moving so fast hardly anyone understood the mission, let alone the equipment. The admiral had given it to her with the look that said, whatever you do don't mess this up. She wasn't accustomed to failure and always seemed to find a way to claw to success, but Kingfisher looked to be a career killer.

Later, Jane flew through the kitchen saying goodbye to her family without even stopping for coffee. She got into her office at the Pentagon early so she could see what headlines were going to blow up her daily schedule. Sure enough, there was mention of possible congressional hearings on the Kingfisher Program. The admiral wanted her in his office at 8:15 sharp. She called for her team to meet at 7:45 for a standup brief (she found that was the only way to keep the team on task and on time, since there was so much to cover on any given day). The standup was testy. It was reported that one contractor (your company), down-selected with another for the design and development phase, was slow in staffing and funding their office and didn't seem energized. In the meeting with the admiral, Jane learned that a feisty senator was zeroing in on Kingfisher as an example of what was "wrong" with the defense acquisition process and she needed to come up with a keep-sold strategy even before program award. The admiral gave her that look again.

Back in her office, Jane gathered her support staff and reorganized her schedule except for critical meetings. It was going to be a long day. If she could just come up with something to show the admiral and the senator that the Kingfisher Program was feasible and on track, maybe she could survive. If she could make it work and it was of real benefit to the fleet, maybe she could get promoted. If only she had someone in her corner. Unfortunately, she didn't think that would be any of the contractors. All they seemed to want to do was show PowerPoint presentations about how great their companies are. At 9 a.m. you walk into her office for your first visit. Now it's up to you. What's your approach? Can you imagine your customer's day? Your customer's life? Can you help?

Life Quotes

"There are only two tragedies in life: one is not getting what one wants, and the other is getting it."
--Oscar Wilde

"If you really want something in this life, you have to work for it – now be quiet, they're about to announce the lottery numbers!"
--Homer Simpson

"The surest sign of intelligent life in the universe is that they haven't attempted to contact us."
--Bill Watterson

"When we remember we are all mad, the mysteries disappear and life stands explained."
--Mark Twain

"My life has no purpose, no direction, no aim, no meaning, and yet I'm happy. I can't figure it out. What am I doing right?"
-- Charles Schulz

"Life is 10 percent what you make it, and 90 percent how you take it."
--Irving Berlin

"Seventy percent of success in life is showing up."
--Woody Allen

"I have a simple philosophy: Fill what's empty. Empty what's full. Scratch where it itches."
--Alice Roosevelt Longworth

"All life is an experiment. The more experiments you make the better."
--Ralph Waldo Emerson

"The truth is you don't know what is going to happen tomorrow. Life is a crazy ride, and nothing is guaranteed."
--Eminem

"Live the life you have imagined."
--Henry David Thoreau

"Life is really simple, but we insist on making it complicated."
--Confucius

"Any idiot can face a crisis - It's day-to-day living that wears you out."
--Anton Chekhov

"I arise in the morning torn between a desire to improve the world and a desire to enjoy the world. This makes it hard to plan the day."
--E. B. White

"Life's but a walking shadow, a poor player, that struts and frets his hour upon the stage, and then is heard no more; it is a tale told by an idiot, full of sound and fury, signifying nothing."
--William Shakespeare

"The art of living is more like wrestling than dancing."
--Marcus Aurelius

"A person will sometimes devote all his life to the development of one part of his body - the wishbone."
--Robert Frost

"Life is a succession of lessons which must be lived to be understood."
--Helen Keller

"Most people have never learned that one of the main aims in life is to enjoy it."
--Samuel Butler

"Life is a tragedy when seen in close-up, but a comedy in long-shot."
--Charlie Chaplin

"Life well spent is long."
--Leonardo da Vinci

"While there's life, there's hope."
--Marcus Tullius Cicero

"Life loves the liver of it."
--Maya Angelou

"Life is pleasant. Death is peaceful. It's the transition that's troublesome."
--Isaac Asimov

"Life isn't a matter of milestones, but of moments."
--Rose Kennedy

"To change one's life: Start immediately. Do it flamboyantly."
--William James

"Life is a moderately good play with a badly written third act."
--Truman Capote

"Life is anything that dies when you stomp on it."
--Dave Barry

"To live is so startling it leaves little time for anything else."
--Emily Dickinson

"We are all here for a spell; get all the good laughs you can."
--Will Rogers

"Variety's the very spice of life, that gives it all its flavor."
--William Cowper

"Life is a lot like jazz... it's best when you improvise."
--George Gershwin

"I believe that if life gives you lemons, you should make lemonade... And try to find somebody whose life has given them vodka."
--Ron White

"Life is one big road with lots of signs. So when you riding through the ruts, don't complicate your mind."
--Bob Marley

"You will get all you want in life if you help enough other people get what they want."
--Zig Ziglar

"Life is tough, but it's tougher when you're stupid."
--John Wayne

"May you live all the days of your life."
--Jonathan Swift

EMOTIONS - EMOTIONAL DECISIONS

Emotions

Are you aware that emotions rule many aspects of the complex and highly technical aerospace, defense, and technology business sectors? For example, our stock prices are subject to *perceptions* of our leadership, talent, and customer base. And whether or not we can convince our customers to buy from us depends a great deal on how well they *like* and *trust* us.

Every business should be concerned about customer satisfaction. It's said that the one true metric of customer satisfaction is whether or not they would refer you to others. You often see that question on customer surveys, because it's a litmus test. If you're willing to refer a company to someone else, you likely are in the upper fifth of that company's customer base. As a *passionate* customer, you can't think of a world without that company and you're willing to give them the majority of your business and recommend them to your friends and business associates.

If a company you like performs well, you take pride in its achievement. This psychological effect, known as B.I.R.G. (basking in reflected glory), was described by relations guru Robert Cialdini (author of *Influence: The Science of Persuasion*) and his colleagues in a study of college sports fans. People associated with winners "basked" in the triumphant spirit, while those associated with losers distanced themselves from their teams. The same holds true for brand companies you deal with. Think of a company you like to buy from. Perhaps a car manufacturer or fashion house. When they're successful it makes you feel good, doesn't it? By doing well, that brand validated your decision to buy from them ("You chose wisely." - Monty Python and the Holy Grail). When they win, you feel like you've won. That's why we instinctively know that mission success and past performance are essential in gaining and keeping government customers.

I believe it's useful to think of customer satisfaction as having two aspects: *technical and emotional*. They're both important, but usually we focus more on the technical part. This helps explain why we can be performing to cost and schedule, but get a less-than-stellar CPAR from the U.S government customer. The funny thing about customer satisfaction is that if a product arrives on time and works perfectly, there still can be mid-range customer attitudes about the provider. There's just not much emotion involved. The product is doing exactly what it's expected it to do. It's become a utility like electrical power. But now let's say something goes wrong. If the provider acknowledges the problem, apologizes for the inconvenience and fixes or replaces the product immediately, their customer satisfaction ratings skyrocket. Emotional trust has been added to the mix. The irony here is that problems, correctly and quickly solved, have the capacity to increase customer satisfaction. Emotions rule.

Emotion Quotes

"The emotions aren't always immediately subject to reason, but they are always immediately subject to action."
--William James

"We make mistakes because the easiest and most comfortable course for us is to seek insight where it accords with our emotions - especially selfish ones."
--Alexander Solzhenitsyn

"Your intellect may be confused, but your emotions will never lie to you."
--Roger Ebert

"The sign of an intelligent people is their ability to control their emotions by the application of reason."
--Marya Mannes

"At the constitutional level where we work, 90 percent of any decision is emotional. The rational part of us supplies the reasons for supporting our predilections."
--William O. Douglas

"I don't know why, it's the same reason why you like some music and you don't like others. There's something about it that you like. Ultimately I

don't find it's in my best interests to try and analyze it, since it's fundamentally emotional."
--Jerry Garcia

"If your emotional abilities aren't in hand, if you don't have self-awareness, if you are not able to manage your distressing emotions, if you can't have empathy and have effective relationships, then no matter how smart you are, you are not going to get very far."
--Daniel Goleman

"Never build your emotional life on the weaknesses of others."
--George Santayana

"People don't buy for logical reasons. They buy for emotional reasons."
--Zig Ziglar

"The first and simplest emotion which we discover in the human mind, is curiosity."
--Edmund Burke

"I do not think there is any thrill that can go through the human heart like that felt by the inventor as he sees some creation of the brain unfolding to success... Such emotions make a man forget food, sleep, friends, love, everything."
--Nikola Tesla

"The finest emotion of which we are capable is the mystic emotion."
--Albert Einstein

"Any emotion, if it is sincere, is involuntary."
--Mark Twain

"When dealing with people, remember you are not dealing with creatures of logic, but creatures of emotion."
--Dale Carnegie

"A work of art which did not begin in emotion is not art."
--Paul Cezanne

"Human behavior flows from three main sources: desire, emotion, and knowledge."
--Plato

"Never trust anyone who wants what you've got. Friend or no, envy is an overwhelming emotion."
--Eubie Blake

"Gratitude isn't a burdening emotion."
--Loretta Young

"A person buying ordinary products in a supermarket is in touch with his deepest emotions."
--John Kenneth Galbraith

"Even though people may be well known, they hold in their hearts the emotions of a simple person for the moments that are the most important of those we know on earth: birth, marriage and death."
--Jackie Kennedy

"Good manners have much to do with the emotions. To make them ring true, one must feel them, not merely exhibit them."
--Amy Vanderbilt

"We know too much and feel too little. At least, we feel too little of those creative emotions from which a good life springs."
--Bertrand Russell

"Let's not forget that the little emotions are the great captains of our lives and we obey them without realizing it."
--Vincent Van Gogh

"Researchers have found that even more than IQ, your emotional awareness and abilities to handle feelings will determine your success and happiness in all walks of life, including family relationships."
--John Gottman

"Compelling reason will never convince blinding emotion."
--Richard Bach

"There can be no transforming of darkness into light and of apathy into movement without emotion."
--Carl Jung

"The heart is the first feature of working minds."
--Frank Lloyd Wright

Emotional Decisions

Emotions can win out over logic in decisions. It can be a surprise when customers don't make what you consider to be logical decisions based on factual comparisons and best choices for the future. Classical decision making is defined as the outcome of mental processes culminating in the selection of a course of action among alternatives. But scientists know that decision making is also an emotional process that can be rational or irrational based on underlying assumptions and feelings.

Scientific studies have shown that many decisions are made in the brain's limbic system (the set of structures connected to emotions) and that the neocortex (the newer, outer brain sector involved with higher functions such as perception and thought) is then tasked to support these emotion-based decisions by coming up with "facts." Of course, these words are a simplification of a very complicated process, but it does help explain why emotion-based messaging (like advertising, for example) can be so convincing.

I don't know if the primacy of emotions in decision making holds true in all cases, but I can relate how it happened to me. Several years ago I was in need of replacing the big television in the family room. Did you have one of those big boys? It took four of us to get it out of there. Since I was making a capital investment I tried to play the part of a sophisticated government key decision maker. I laid out my specifications and set a budget. I even had two down-select candidates. During my "look-off" competition at Best Buy, things were going swimmingly until I looked over and became enraptured with a particular HD flat-screen model. I still don't know what happened, but suddenly I was standing in front of those beautiful colors and I fell in love. My limbic system apparently ordered my neocortex to come up with facts and my brain went into overdrive to justify this television and blow my budget by $500. It had more HDMI ports, a higher refresh rate and easy sync with the new Blu-ray/DVD player I wanted. I still love that television. I come home at night and hug it. That television holds both my technical and emotional satisfaction.

Customer decisions are affected by a variety of other factors, such as personality types, generational groupings, and organizational cultures, but it is the emotional part that remains most difficult for scientists, engineers, and IT professionals to understand in the decision process. They have a hard time comprehending that a key decision maker's emotion-based "personal values" can be just as important as financial and performance data in deciding who gets the contract. Feelings are facts and you need to account for them when convincing customers to keep choosing you.

Decision Quotes

"When confronted with two courses of action, I jot down on a piece of paper all the arguments in favor of each one, then on the opposite side I write the arguments against each one. Then by weighing the arguments pro and con and cancelling them out, one against the other, I take the course indicated by what remains."
--Benjamin Franklin

"He who has a choice has trouble."
--Dutch proverb

"Alternatives, and particularly desirable alternatives, grow only on imaginary trees."
--Saul Bellow

"Between two stools one sits on the ground."
--French proverb

"The absence of alternatives clears the mind marvelously."
--Henry Kissinger

"You can't build a strong corporation with a lot of committees and a board that has to be consulted every turn. You have to be able to make decisions on your own."
--Rupert Murdoch

"Every decision you make is an important one, whether there are twenty thousand people working for you or just one."
--Donald Trump

"Be willing to make decisions. That's the most important quality in a good leader."
--General George S. Patton

"Choose always the way that seems the best, however rough it may be; custom will soon render it easy and agreeable."
--Pythagoras

"The most difficult thing is the decision to act, the rest is merely tenacity."
--Amelia Earhart

"All my life, whenever it comes time to make a decision, I make it and forget about it."
--Harry S Truman

"Decide which is the line of conduct that presents the fewest drawbacks and then follow it out as being the best one, because one never finds anything perfectly pure and unmixed, or exempt from danger."
--Niccolo Machiavelli

"The best we can do is size up the chances, calculate the risks involved, estimate our ability to deal with them, and then make our plans with confidence."
--Henry Ford

"I think we should follow a simple rule: if we can take the worst, take the risk."
--Dr. Joyce Brothers

"To decide against your conviction is to be an unqualified and inexcusable traitor, both to yourself and to your country, let men label you as they may."
--Mark Twain

"A wise man makes his own decisions, an ignorant man follows public opinion."
--Chinese Proverb

"Every man must decide whether he will walk in the light of creative altruism or in the darkness of destructive selfishness."
--Martin Luther King, Jr.

"Sometimes you make the right decision, sometimes you make the decision right."
--Dr. Phil

"I was a good amateur but only an average professional. I soon realized that there was a limit to how far I could rise in the music business, so I left the band and enrolled at New York University."
--Alan Greenspan

"I felt uneasy about making the rapid decisions I have always made."
--Richard Branson

"Quick decisions are unsafe decisions."
--Sophocles

"I have always found that if I move with 75 percent or more of the facts I usually never regret it. It's the guys who wait to have everything perfect that drive you crazy."
--Lee Iacocca

"The opportunity is often lost by deliberating."
--Publilius Syrus

"I truly understand that there is a lesson in everything that happens to us. So I tried not to spend my time asking, "Why did this happen to me?" but trying to figure out why I had chosen this."
--Oprah Winfrey

"Most discussions of decision making assume that only senior executives make decisions or that only senior executives' decisions matter. This is a dangerous mistake."
--Peter F. Drucker

"When making a decision of minor importance, I have always found it advantageous to consider all the pros and cons. In vital matters, however, such as the choice of a mate or a profession, the decision should come from the unconscious, from somewhere within ourselves."
--Sigmund Freud

"A good decision is based on knowledge and not on numbers."
--Plato

"Statistics are no substitute for judgment."
--Henry Clay

"Once the 'what' is decided, the 'how' always follows. We must not make the 'how' an excuse for not facing and accepting the 'what'."
--Pearl S. Buck

"Life is the art of drawing sufficient conclusions from insufficient premises."
--Samuel Butler

"Man is a reasoning, rather than a reasonable, animal."
--Alexander Hamilton

"It is the heart always that sees before the head can see."
--Thomas Carlyle

"Better to be without logic than without feeling."
--Charlotte Bronte

"After a battle is over people talk a lot about how decisions were methodically reached, but actually there's always a hell of a lot of groping around."
--Admiral Frank Jack Fletcher

"Nothing is more difficult, and therefore more precious, than to be able to decide."
--Napoleon Bonaparte

FEAR - FIND A PROBLEM-FIX A PROBLEM - FOCUS ON THE CUSTOMER

Fear

In business development you have to deal with fear. There's plenty to go around: financial crises, natural disasters, and Washington press conferences. They're all scary.

In relating to customers it's important to employ positive emotion to engender trust. But what do you do in an atmosphere of negative emotion, when your customers are fretting over budgetary pressures and you're concerned over your business growth potential? First, you need to confront your own fears. According to Asher Strategies (who run a fine course compatible with government sales), there are basically two fears facing business developers. Number one is the fear of their own lack of knowledge of what they are selling. Recall Dr. Tom Barrett's advice encouraging you to be an ACE when delivering customer messages. You need Authority, Conviction and Enthusiasm to be persuasive. The conviction part comes from knowing your stuff better than anyone else. Knowledge builds your expertise and increases your persuasiveness. Dale Carnegie noted that top salespeople are "knowledge giants," who also know the competition's business better than the competition does. And if you need backup assurance, take along a technical expert (sales engineer) when you make that critical call on the customer.

Sales fear number two is fear of rejection by the prospective customer. Many people drawn to sales are pro-active and people-oriented. They invest a lot of their self-image into what they do and when the customer says no it can be personally devastating. There are a number of mental strategies great salespeople employ to overcome fear of further sales rejections. They understand that many times a loss has nothing to do with them, that many factors are in play. Other times it wasn't a "no" it was just "not now" and if they press on they might be able to bring that customer to closure. By remaining positive in your outlook you can position yourself to make the sale when the customer finally understands the value of your solution or when the competitor they first selected can't perform the work. If you're in high opportunity sectors, such as IT task orders, you need to remind yourself that much of selling is a numbers game. It's very difficult to win every competition. And if you're in high-stakes billion-dollar competitions, you need to recognize that political variables can weigh more heavily on evaluations than price and performance. These are times when your internal best practices (policies and procedures) can kick in and help you focus on high probability sales and what you need to do to ensure sales success.

Customers also have fears. Aside from their own fear of failure (*Will this decision kill me?*), the customer's main buying fear is that they do not fully understand the value of your offering. Those of us who have been government customers in our past lives recall numerous contractor presentations we sat through thunder-struck at our inability to comprehend what was on the charts and what was being said. It might have been the key solution we were seeking or just hand-waving by contractors eager to make a sale. We couldn't tell. Great business developers help customers overcome this fear by employing multiple communication channels. All of us are inundated by information and have difficulty recalling and assembling the most important bits of knowledge. Asher Strategies observes that, generally speaking, we have concept retention of about 10 percent of what we read, 12 percent of what we hear and 30 percent of what we see. Our recall rate jumps to 50 percent when we've watched a film clip. In many cases we're able to recall 70 percent of information when we've participated in lively back-and-forth discussions. But the real shocker is that there's potential to recall 90 percent of critical information if we actually have done the activity ourselves, or simulated the real thing or participated in a pilot program. So the more you can get those key decision makers physically and emotionally involved with your key programs, the more potential they have to fully understand the value of what your solutions can do for them and the more they can put aside their fears and trust you.

So put on your No Fear T-Shirt and go out there and sell something.

Fear Quotes

"The only thing we have to fear is fear itself."
--Franklin D Roosevelt

"Nothing in life is to be feared. It is only to be understood."
--Marie Curie

"Keep your fears to yourself but share your courage with others."
--Robert Louis Stevenson

"There is a time to take counsel of your fears, and there is a time to never listen to any fear."
--George S. Patton

"To conquer fear is the beginning of wisdom."
--Bertrand Russell

"Fear makes us feel our humanity."
--Benjamin Disraeli

"Fear is static that prevents me from hearing myself."
--Samuel Butler

"Fear makes the wolf bigger than he is."
--German Proverb

"A cat bitten once by a snake dreads even rope."
--Arab Proverb

"Worry gives a small thing a big shadow."
--Swedish proverb

"I have learned over the years that when one's mind is made up, this diminishes fear; knowing what must be done does away with fear."
--Rosa Parks

"Don't be afraid to go out on a limb. That's where the fruit is."
--H. Jackson Browne

"Have no fear of perfection--you'll never reach it."
--Salvador Dali

"Inaction breeds doubt and fear. Action breeds confidence and courage. If you want to conquer fear, do not sit home and think about it. Go out and get busy."
--Dale Carnegie

"Fear has its use but cowardice has none."
--Mahatma Gandhi

"The way you overcome shyness is to become so wrapped up in something that you forget to be afraid."
--Lady Bird Johnson

"Each time we face our fear, we gain strength, courage, and confidence in the doing."
--Theodore Roosevelt

"Fear is your best friend or your worst enemy."
--Mike Tyson

"Better beans and bacon in peace than cakes and ale in fear."
--Aesop

"There are two levers for moving men: interest and fear."
--Napoleon Bonaparte

"Nothing is more despicable than respect based on fear."
--Albert Camus

"There is no feeling, except the extremes of fear and grief, that does not find relief in music."
--George Eliot

"Ignorance is the mother of fear."
--Eric Hoffer

"Fear not, provided you fear; but if you fear not, then fear."
--Blaise Pascal

"Never let the fear of striking out get in your way."
--Babe Ruth

"Courage is resistance to fear, mastery of fear - not absence of fear."
--Mark Twain

"The man who fears no truths has nothing to fear from lies."
--Sir Francis Bacon

"He who is not everyday conquering some fear has not learned the secret of life."
--Ralph Waldo Emerson

"To him who is in fear everything rustles."
--Sophocles

"In this world there is always danger for those who are afraid of it."
--George Bernard Shaw

"Better hazard once than always be in fear."
--Thomas Fuller

"When even one American -- who has done nothing wrong -- is forced by fear to shut his mind and close his mouth -- then all Americans are in peril."
--Harry S Truman

"Avoiding danger is no safer in the long run than outright exposure. The fearful are caught as often as the bold."
--Helen Keller

"The fearful unbelief is unbelief in yourself."
--Thomas Carlyle

"With the fearful strain that is on me night and day, if I did not laugh I should die."
--Abraham Lincoln

"To me, Fearless is living in spite of those things that scare you to death."
--Taylor Swift

"Keep fear alive."
--Stephen Colbert

Find a Problem, Fix a Problem

A good way to increase customer trust is to find a problem and fix it. During customer engagements, business developers and program managers usually wait for the customer to say if anything's wrong. If something particular is really bothering a customer, they'll let you know, right? Not necessarily. TARP Worldwide conducted original research on customer complaints and found that, on average, 50 percent of customers will complain to a front-line person, but no more than 10 percent will complain to a manager. Think about when you had a problem with a company and were mad. Sometimes you didn't want the hassle and just left, right? In working with angry government customers, they not only can leave without warning, they also can hurt your business elsewhere. This suggests you have work to do in improving customer trust even though your customer doesn't say so.

Where to start? Research also has shown that dissatisfied customers eventually tell at least nine other people of their problems face-to-face. These could be in the customer set or in related organizations. Good customer conversations extend beyond your direct user customer to the entire customer chain above and below, such as administrative buyers and subcontractors. Someone somewhere knows something about your customer's satisfaction with your products and services. It takes time and effort to find undisclosed problems.

A good place to look for a problem irritating customers is in the area of processes. In the aerospace, defense, and technology business sectors, processes are everywhere. Many are necessary for legal reasons and to achieve mission success. For example, to fix a customer problem you can't break regulations related to public safety and protection or underlying statutory laws like the Federal Acquisition Regulation. But maybe you can adjust or eliminate irritating guidelines - procedures designed to serve your organization, inefficient internal processes, and policies crafted for convenience. Just because your group has always done something the same way shouldn't rule out beneficial change. A good time to implement change is when there's an imperative to streamline your organizations. Altering (or eliminating) a process that irritates both you and your customer can be a solution to two problems at once: your need to reduce overhead and increase customer satisfaction.

Unless you need specific information to start, don't wait for permission from the customer to fix a problem. Let them worry about other problems with other contractors. Have you ever worked with a great contractor or service representative who was one step ahead of you, anticipating what you needed even before you thought about it? Sometimes their fixes were no cost and sometimes they were cost options that could save you money in the long run. I had a service representative once call me and say that after a review of my account they were reducing my monthly charge for the same service. I'm still with that company, even though I've been promised "better for cheaper" by competitors.

If you want to rapidly move the needle up on your customer's emotional satisfaction, find and fix a problem. Make it go away without them asking. Problems can be viewed as opportunities to increase customer satisfaction and trust, turning wary customers into loyal customers. Just a small increase in customer loyalty can return large profits. And loyal customers stay with you in hard times.

Problem Quotes

"The significant problems we face cannot be solved at the same level of thinking we were at when we created them."
--Albert Einstein

"Our problems are man-made, therefore they may be solved by man."
--John F. Kennedy

"I'd say it's been my biggest problem all my life... it's money. It takes a lot of money to make these dreams come true."
--Walt Disney

"A problem is a chance for you to do your best."
--Duke Ellington

"For every complex problem there is an answer that is clear, simple, and wrong."
--H. L. Mencken

"If you only have a hammer, you tend to see every problem as a nail."
--Abraham Maslow

"The government solution to a problem is usually as bad as the problem."
--Milton Friedman

"The problem with having a sense of humor is often that people you use it on aren't in a very good mood."
--Lou Holtz

"No problem can withstand the assault of sustained thinking."
--Voltaire

"It is no good getting furious if you get stuck. What I do is keep thinking about the problem but work on something else. Sometimes it is years before I see the way forward. In the case of information loss and black holes, it was 29 years."
--Stephen Hawking

"It is a common experience that a problem difficult at night is resolved in the morning after the committee of sleep has worked on it."
--John Steinbeck

"I have found that the greatest help in meeting any problem is to know where you yourself stand. That is, to have in words what you believe and are acting from."
--William Faulkner

"Doing what's right isn't the problem. It is knowing what's right."
--Lyndon B. Johnson

"Don't bother people for help without first trying to solve the problem yourself."
--Colin Powell

"Lack of knowledge... that is the problem."
--W. Edwards Deming

"Each problem that I solved became a rule, which served afterwards to solve other problems."
--Rene Descartes

"The problem with people who have no vices is that generally you can be pretty sure they're going to have some pretty annoying virtues."
--Elizabeth Taylor

"The whole problem with the world is that fools and fanatics are always so certain of themselves, and wiser people so full of doubts."
--Bertrand Russell

"Indecision may or may not be my problem."
--Jimmy Buffett

"A problem well stated is a problem half solved."
--John Dewey

"When I am working on a problem, I never think about beauty but when I have finished, if the solution is not beautiful, I know it is wrong."
--R. Buckminster Fuller

"Each success only buys an admission ticket to a more difficult problem."
--Henry A. Kissinger

"The measure of success is not whether you have a tough problem to deal with, but whether it is the same problem you had last year."
--John Foster Dulles

"Some problems are so complex that you have to be highly intelligent and well-informed just to be undecided about them."
--Laurence J. Peter

"Problems are only opportunities in work clothes."
--Henri J. Kaiser

"My problem with chess was that all my pieces wanted to end the game as soon as possible."
--Dave Barry

"The most serious mistakes are not being made as a result of wrong answers. The truly dangerous thing is asking the wrong questions."
--Peter Drucker

"Our major obligation is not to mistake slogans for solutions."
--Edward R. Murrow

"Don't find fault, find a remedy."
--Henry Ford

"If you can talk brilliantly about a problem, it can create the consoling illusion that it has been mastered."
--Stanley Kubrick

"The only difference between a problem and a solution is that people understand the solution."
--Charles Kettering

"You know what your problem is, it's that you haven't seen enough movies – all of life's riddles are answered in the movies."
--Steve Martin

Focus on the Customer

Large account aerospace, defense and technology customers can easily become grouchy, reflecting the pressures they're feeling from all directions. You can't help them solve every problem. But you can do your best to be easy to work with, deliver on your programs, and focus on your customers. Here's a customer focus checklist developed by Bob Trice, who inspired several generations of business developers at McDonnell Douglas, General Dynamics, and Lockheed Martin. The checklist can also apply to program managers who work face-to-face with customers and many others throughout your organization, including those who have internal customers. Take a look and see if there's something here you might want to consider the next time you connect with your customer.

CUSTOMER FOCUS CHECKLIST

- *Who, exactly, are my most important customers?*
- *What are my customers' key concerns?*
- *How am I and my organization addressing my customers' problems?*
- *When did I last personally meet with my key customers?*
- *What issues were discussed with each customer?*
- *Did I take any actions out of these meetings? Have I completed these actions and reported back to the customer?*
- *Did I distribute feedback from these meetings to the right people in my organization and others that may be affected within the company?*
- *Who else within the company shares my key customers? Have I coordinated with them before seeing the customer?*

- *Who within the company should be working at levels above and below my customers to reinforce our company position/message?*
- *How frequently do I need to visit my customers? Do I need to coordinate with the field offices?*
- *When will I next visit my customer set?*

Focus Quotes

"Be dramatically willing to focus on the customer at all costs, even at the cost of obsoleting your own stuff."
--Scott Cook

"You can't depend on your eyes when your imagination is out of focus."
--Mark Twain

"That's been one of my mantras – focus and simplicity. Simple can be harder than complex: You have to work hard to get your thinking clean to make it simple. But it's worth it in the end because once you get there, you can move mountains."
--Steve Jobs

"Concentrate all your thoughts upon the work at hand. The sun's rays do not burn until brought to a focus."
--Alexander Graham Bell

"My focus is to forget the pain of life. Forget the pain, mock the pain, reduce it. And laugh."
--Jim Carrey

"When words become unclear, I shall focus with photographs."
--Ansel Adams

"The game has its ups and downs, but you can never lose focus of your individual goals and you can't let yourself be beat because of lack of effort."
--Michael Jordan

"Shoot a few scenes out of focus. I want to win the foreign film award."
--Billy Wilder

"I'm happiest when I have something to focus my energy on."
--Scarlett Johansson

"You shouldn't focus on why you can't do something, which is what most people do. You should focus on why perhaps you can, and be one of the exceptions."
--Steve Case

"To create something exceptional, your mindset must be relentlessly focused on the smallest detail."
--Giorgio Armani

"I believe that as much as you take, you have to give back. It's important not to focus on yourself too much."
--Nicole Kidman

"Scientists will eventually stop flailing around with solar power and focus their efforts on harnessing the only truly unlimited source of energy on the planet: stupidity. I predict that in the future, scientists will learn how to convert stupidity into clean fuel."
--Scott Adams

"Focus on what you have."
--Suze Orman

"Sometimes in the past when I played something might make me lose focus, or I would go home after a game where I thought I could have played better and I would let it hang over my head for a long time when it shouldn't."
--LeBron James

"Focus on remedies, not faults."
--Jack Nicklaus

"It is very difficult to get people to focus on the most important things when you're in boom times."
--Jeff Bezos

"Beware lest you lose the substance by grasping at the shadow."
--Aesop

"A man should never be appointed into a managerial position if his vision focuses on people's weaknesses rather than on their strengths."
--Peter Drucker

"I don't care how much power, brilliance or energy you have, if you don't harness it and focus it on a specific target, and hold it there you're never going to accomplish as much as your ability warrants."
--Zig Ziglar

"My success, part of it certainly, is that I have focused in on a few things."
--Bill Gates

"The successful warrior is the average man, with laser-like focus."
--Bruce Lee

"A corporation is a living organism; it has to continue to shed its skin. Methods have to change. Focus has to change."
--Andrew Grove

"A person who aims at nothing is sure to hit it."
--Anonymous

"A major stimulant to creative thinking is focused questions. There is something about a well-worded question that often penetrates to the heart of the matter and triggers new ideas and insights."
--Brian Tracy

"As you become more clear about who you really are, you'll be better able to decide what is best for you – the first time around."
--Oprah Winfrey

"Champions get what they want because they know what they want. They have a vision that keeps them motivated and efficiently on track. They see it, feel it, and experience it in their minds and hearts. What is success for you? You won't get there without knowing what it feels and looks like."
--Dr. Phil

"One very important aspect of motivation is the willingness to stop and to look at things that no one else has bothered to look at. This simple process of focusing on things that are normally taken for granted is a powerful

source of creativity."
--Edward de Bono

"The only way to advertise is by not focusing on the product."
--Calvin Klein

"Ultimately, we wanted Nike to be the world's best sports and fitness company. Once you say that, you have a focus. You don't end up making wing tips or sponsoring the next Rolling Stones world tour."
--Phil Knight

"When walking, walk. When eating, eat."
--Zen Proverb

GENERATIONS - GOLF AND LIFE AND BUSINESS

Generations

There's common agreement that people born during the same generation share many common characteristics. The main generational divisions currently recognized are:

- Traditionalists - Born 1945 and before
- Boomers - Born 1946 to 1964
- Generation X - Born 1965-1980
- Millennials - Born after 1980

[Notes: Birth dates are approximate. Traditionalists have also been called Veterans, Builders, and the Silent Generation; Boomers were originally referred to as Baby Boomers; Generation Xers were at first called Baby Busters; and the Millennials are also called Generation Y and the Nexters.]

You may have noticed and discussed generations in your workforce. But have you also noticed changes in the generations of your customers? As you relate to your different generational customers, what should you consider when communicating with them?

Here are some helpful observations:

- Traditionalists have zeal toward hard work and dedication to their organizations and mostly communicate formally through proper procedures and channels. Don't call on them without purpose.
- Boomers have a more youthful optimism, though they also respect power and the accomplishment of others. They also communicate somewhat formally through structured networks, but are more relaxed than the previous generation. Call them just to chat sometimes.

- Gen Xers are known to be pragmatic and self-reliant, much more similar to their Traditionalist grandparents, but they value work-life balance and will take Friday off and shut down their cell phones if given the opportunity. They prefer to communicate more casually and directly and will use an e-mail or text message to set up what to them is a more formal phone conversation.
- Millennials work fast and play hard. They don't have distinct boundaries between work and the rest of their lives, but they do have seemingly unbounded confidence. They would like the freedom to work on a project Friday night after work and take off Monday morning. They are pleasers and communicate quickly and casually.

So how should you approach various customer generations? Here are some "customer tips" adapted from *Generations at Work* by Claire Ron and Bob Filipczak:

For Traditionalists

- Establish rapport the traditional way: Mr., Mrs., Sir, Ma'am, Please, Thank You, etc. with good grammar and clear enunciation.
- Set a relaxed pace. Do not rush things. Work through your conversations in a linear order.
- Keep a respectful "distance" between yourself and your Traditionalist customers and avoid being overly chummy or too personal.

For Boomers

- Be personable in greeting and use first names when appropriate. Most Boomers appreciate name recognition.
- Treat them like friends and take time to check in and ask them how they are doing. Talk about all things interesting.
- If they are a continuing customer, help them feel it's a "special" relationship.

For Gen Xers

- Be prompt and efficient, as competence is more appealing to Gen X customers than schmoozing.

- Know your facts, get ready to answer a lot of questions and be ready to share information with a skeptical audience.
- Give distance and allow them to make their own decisions. Don't hover.

For Millennials
- In dealing with younger customers, try not to treat them all the same. There can be as much generational difference between Millennials and Gen Xers as between Gen Xers and the Boomers.
- Don't give the impression you're talking down to them. Millennial customers are young, but also smart and have huge potential. Ask about their interests.
- Quicken the pace as they're accustomed to being entertained and find methodical people and long waiting times boring.

Those are great tips, but it's a lot to remember. One easy way I use to gain insight into how the generations function and how best to communicate with them is to compare the "mass media" of their age. What electronic entertainment did they grow up with? Traditionalists grew up with the radio. Have you ever listened to the radio classics station on SiriusXM? Many of those old radio shows were captivating, but the story lines unfolded in a stately, linear fashion. That's the way you want to reach the Traditionalist's "inner radio." *Keep it safe.* Give them logic, and respect and avoid surprises of any kind.

Boomers grew up with the television. They extended their emotional experiences through dramas, comedies, and game shows shown on that new electronic member of the family. Boomers love to be included and have less emotional distance than Traditionalists. *Keep it real, man.* Give them connection and pleasant surprises.

Gen Xers grew up with video games. They had fun in closed systems of input and feedback. Distractions were, well, distracting. So don't push them. They're smart and will get back to you when they're ready. *Keep it factual.* Give them time to decide.

Millennials grew up with computers and everything connected to everything. They are 24/7 and rarely turn off their cell phones. Hugely talented and with great potential, they struggle with data overload. Help them sort through the data and build long-term views (strategies). *Keep it lively.* Not infotainment, but fast-paced conversations and presentations. Above all, give them quality feedback.

I hope the above will help you better relate to customers from differing generations. It doesn't come easy. There's a lesson I've painfully learned and that is how easy it is to fall into the trap of being "dad" or "mom" to a younger generation customer. We all need to respect the talents of the generations coming behind us and build relationships with them where we can and offer useful guidance, if asked. And for the younger generations dealing with the older ones, it's equally easy to fall into the trap of appearing lightweight and inconsequential if you text during meetings and ask "How am I doing?" too often. It's all about finding the best ways to communicate. While customers are not all the same, almost every customer will respond positively if you are respectful, friendly, attentive, interested and responsive. It's just basic human nature.

Generation Quotes

"I see no hope for the future of our people if they are dependent on the frivolous youth of today, for certainly all youth are reckless beyond words."
--Hesiod (700bc)

"Every generation needs a new revolution."
--Thomas Jefferson

"You will never know how much it has cost my generation to preserve your freedom. I hope you will make good use of it."
--John Quincy Adams

"The philosophy of the school room in one generation will be the philosophy of government in the next."
--Abraham Lincoln

"There is a mysterious cycle in human events. To some generations much is given. Of other generations much is expected. This generation of Americans has a rendezvous with destiny."
--Franklin D. Roosevelt

"No member of our generation who wasn't a Communist or a dropout in the thirties is worth a damn."
--Lyndon B. Johnson

"My generation, faced as it grew with a choice between religious belief and existential despair, chose marijuana. Now we are in our Cabernet stage."
--Peggy Noonan

"We're the middle children of history.... no purpose or place. We have no Great War, no Great Depression. Our great war is a spiritual war. Our great depression is our lives."
--*Fight Club* (Referring to Generation X)

"I think that romance has been taken away a bit for my generation."
--Emma Watson

"It's hard for me to get used to these changing times. I can remember when the air was clean and sex was dirty."
--George Burns

"The first half of our lives is ruined by our parents and the second half by our children."
--Clarence Darrow

"Small children disturb your sleep, big children your life."
--Yiddish Proverb

"There was no respect for youth when I was young, and now that I am old, there is no respect for age - I missed it coming and going."
--J.B. Priestly

"Generation Gap: A chasm, amorphously situated in time and space, that separates those who have grown up absurd from those who will, with luck, grow up absurd."
--Bernard Rosenberg

"They say genes skip generations. Maybe that's why grandparents find their grandchildren so likeable."
--Joan McIntosh

"The older generation thought nothing of getting up at five every morning - and the younger generation doesn't think much of it either."
--John J. Welsh

"First we are children to our parents, then parents to our children, then parents to our parents, then children to our children."
--Milton Greenblatt

"Parents often talk about the younger generation as if they didn't have anything to do with it."
--Haim Ginott

"Our generation has an incredible amount of realism, yet at the same time it loves to complain and not really change. Because, if it does change, then it won't have anything to complain about."
--Tori Amos

"That which seems the height of absurdity in one generation often becomes the height of wisdom in another."
--Adlai Stevenson

"Each generation wants new symbols, new people, new names. They want to divorce themselves from their predecessors."
--Jim Morrison

"Every generation laughs at the old fashions, but follows religiously the new."
--Henry David Thoreau

"I've led this empty life for over forty years and now I can pass that heritage on and ensure that the misery will continue for at least one more generation."
--Larry David

"An author ought to write for the youth of his own generation, the critics of the next, and the schoolmaster of ever afterwards."
--F. Scott Fitzgerald

"Back, you know, a few generations ago, people didn't have a way to share information and express their opinions efficiently to a lot of people. But

now they do."
--Mark Zuckerberg

"The year I was born, 1956, was the peak year for babies being born, and there are more people essentially our age than anybody else. We could crush these new generations if we decided too."
--Tom Hanks

"There are generations who watch *Doctor Who* together."
--Sarah Sutton

"Whole generations have forgotten history."
--Pierre Salinger

"If I were given the opportunity to present a gift to the next generation, it would be the ability for each individual to learn to laugh at himself."
--Charles M. Schulz

Golf and Life and Business

Not long ago a friend greeted me at a business meeting and said he really liked my customer relations observations, but thought my subject matter was getting a little too serious. Now, those of you who know me understand that I have rarely been accused of being too serious, so I took his advice to heart and began to reflect on it. I recalled Dr. Tom Barrett's reply in one of our customer relations classes to the question, "What is the one thing we could do to improve our business?" Without hesitation Tom said: "Lighten up. Take your programs seriously, but don't take yourselves too seriously."

That's when I began to think about golf. I'm not a frequent player, but I've always admired the game for three reasons. The first is that it gets you outside and reconnected to nature. You can reset your life and put things more in balance on a beautiful morning along the fairways. The second reason is that golf allows you to test yourself against your own personal goals as well as opponents. Plus, achievement in golf is relative to how well you can handle both the physical and emotional aspects of the game. And the third reason is that golf can have a wonderful socializing function. It's a very inclusive sport. In business, a round of golf with clients and customers (whenever the lawyers allow it) can help build friendships, partnerships, and teams for life.

Hmmm. That still sounds pretty serious, so here are some classic golf quotes:

Golf Quotes

"To find a man's true character, play golf with him."
--P.G. Wodehouse

"The only sure rule in golf is - he who has the fastest cart never has to play the bad lie."
--Mickey Mantle

"I don't fear death, but I sure don't like those three-footers for par."
--Chi Chi Rodriguez

"Swing hard in case you hit it."
--Dan Marino

"My favorite shots are the practice swing and the conceded putt. The rest can never be mastered."
--Lord Robertson

"There is no similarity between golf and putting; they are two different games, one played in the air, and the other on the ground."
--Ben Hogan

"To play well you have to have good balance in your life."
--Annika Sorenstam

"Professional golf is the only sport where, if you win 20% of the time, you're the best."
--Jack Nicklaus

"If you watch a game, it's fun. If you play at it, it's recreation. If you work at it, it's golf."
--Bob Hope

"If you think it's hard to meet new people, try picking up the wrong golf ball."
--Jack Lemmon

"My swing is so bad, I look like a caveman killing his lunch."
--Lee Trevino

"I'm working as hard as I can to get my life and my cash to run out at the same time. If I can just die after lunch Tuesday, everything will be perfect."
--Doug Sanders

"Long ago when men cursed and beat the ground with sticks, it was called witchcraft. Today it's called golf."
--Will Rogers

"Golf is a good walk spoiled."
--Mark Twain

"It took me seventeen years to get three thousand hits in baseball. It took one afternoon on the golf course."
--Hank Aaron

"Golf is a game whose aim is to hit a very small ball into an ever smaller hole, with weapons singularly ill-designed for the purpose."
--Winston Churchill

"I know I am getting better at golf because I am hitting fewer spectators."
--Gerald R. Ford

"Golf: A plague invented by the Calvinistic Scots as a punishment for man's sins."
--James Barrett Reston

"Although golf was originally restricted to wealthy, overweight Protestants, today it's open to anybody who owns hideous clothing."
--Dave Barry

"I asked the Dalai Lama the most important question, I think, that you could ask him — if he'd ever seen Caddyshack."
--Jesse Ventura

"Golf is a game of coordination, rhythm and grace; women have these to a high degree."
--Babe Didrikson Zaharias

"The most important shot in golf is the next one."
--Ben Hogan

"I have a tip that can take five strokes off anyone's game: It's called an eraser."
--Arnold Palmer

"Golf is a puzzle without an answer. I've played the game for 50 years and I still haven't the slightest idea of how to play."
--Gary Player

"Golf appeals to the idiot in us and the child. Just how childlike golf players become is proven by their frequent inability to count past five."
--John Updike

"We learn so many things from golf—how to suffer, for instance."
--Bruce Lansky

"The reason a pro tells you to keep your head down is so you can't see him laughing."
--Phyllis Diller

"Golf combines two favorite American pastimes; taking long walks and hitting things with a stick."
--P.J. O'Rourke

"Golf is a game in which you yell 'fore', shoot six and write down five."
--Paul Harvey

"Golf is not, on the whole, a game for realists. By its exactitude's of measurements, it invites the attention of perfectionists."
--Heywood Hale Broun

"Golf's three ugliest words: Still your shot."
--Dave Marr

"I never learned anything from a match that I won."
--Bobby Jones

"The worst club in my bag is my brain."
--Chris Perry

HANDLING DIFFICULT CUSTOMERS - HANDSHAKES - SELLING TO HUMANS

Handling Difficult Customers

I've often been asked, "How do you handle a difficult government customer?" The answer is "Like any other difficult customer." Just because someone works in the public sector and is a manager or administrator of a large program doesn't mean they're all that different when it comes to getting along. Government officials are, however, faced with a lot more stresses than we're usually aware of, brought on by great responsibility and less control over their environments than they would wish.

Many in business development and sales positions aren't familiar with the principles of fixing grumpy customers. Sometimes these customers are having troubles in their personal lives that are carrying over into their professional work. In these cases you might perceive that something's bothering them personally and try not to make their situations any more difficult than they already are. You can't fix a customer's personal problems. But you can be understanding and non-confrontational, while moving toward work-around solutions more agreeable to the customer and your company.

In some cases there are program problems exacerbating underlying tensions that spill out into the open. In some cases it's the customer's fault. Perhaps they haven't managed the program well or haven't adequately informed their leadership of major changes or delays. It's easy to pull back and try to let the blame go to them. But does that really help you and your program? The main objective should be to make the customer look good, to let them be the hero. If you can quietly and calmly help them, you win in the long run.

In other cases a problem with a program is your company's fault. With advanced technologies and system integration complexities, any number of things can go wrong. How do most companies usually handle such issues? The customer confronts them and they push back, deny fault (or go silent), and study the technical details until they can show the customer is wrong. But that just makes the customer more upset and doesn't really fix the emotional part of the problem. Companies resist saying or implying they're sorry. They're cautioned by counsel that saying "sorry" has legal implications. But there are lots of ways to get this message across, avoid incurring additional contractual obligations, and restore trust in the relationship. If you can acknowledge the problem, show you're concerned and vow to fix it quickly, that goes a long way in calming down an irate customer.

I believe every business development or sales professional should be aware of how front line customer service people handle difficult customers. You never know when you might be confronted. If you don't know what to do, you'll likely make a mess of things. Years ago I took a one-day course for call center people put on by Padgett-Thompson. It was a low-cost, high-payoff investment. The following are some of the principles I learned and have applied in many of my own customer situations.

Up to 80 percent of customer-client problems involve poor communication, rather than poor product performance. Many of these are poorly managed customer expectations that escalated into full-blown problems. Considering it costs up to five times as much money to get a new customer as it does to save one, there's good ROI in doing all you can to save a customer. An irate customer, no matter whether angered by a broken toaster or a broken warship, wants the four F's from company representatives: Friendly, Flexible, Fixer, and Follow-Up. Don't forget the follow-up.

In order to fix customers, first you have to understand them and then you have to help them control their emotion. And you must do these actions in that order. If you try and fix them before you know what they need and while they're still emotional, you will fail. So what can help? You can gain understanding of a customer by paying attention to their body movements (see "X-Ray Vision") and how they're speaking. Body language can be up to 55 percent of messaging. If they're tense and agitated, your customers will display their discomfort in their posture and limbs. If you want to calm them down, you must relax and override your own instincts to equally tense up. If you can't control your body, your body will control your message.

When irate customers speak, they tend to do so at a fast rate and high pitch, often slurring words. Their voices may even crack (a red flag if you're in arm's reach). You can help an irate customer communicate by keeping calm and slowing your own rate of speech and pitch (though not too slow and low). Try and match their normal rate of speech - people tend to listen at the same rate they speak. If you are dealing with a non-native English speaker and have trouble understanding their accent, they likely have trouble understanding you. Use appropriate word sets: technical words for technical buyers, executive words for executive buyers.

You can help an irate customer diffuse emotion by letting them vent. It's natural to try and stop them, but often it's better to let the storm play out. If they're in front of an audience, find an opportunity to move them "off stage" where the storm will subside quicker. Focus on the real problem, not on all the other stuff that gets dragged up in venting. Ask questions to elicit not only the problem, but their desired solution. If the storm continues, ask them directly: "What do you want?" Do they have something specific in mind or are they just expressing general frustration? It's also helpful to adjust to a customer's generation and personality types (see "Generations" and "Personalities").

If your customer has a specific fix in mind that you can't accommodate, what do you do? It's hard to tell a customer, "No," but sometimes you have to. You can say "No" positively by stating why you can't say "Yes," such as legal and ethical concerns and the space-time continuum. Then, and it's a matter of timing, quickly tell them what you *can* do. If you don't know what you can do at the moment to fix a customer, you have to ask yourself if you are taking full responsibility for your job. If your boss doesn't give you authority to fix a customer, ask for it. And remember, sometimes saying yes doesn't result in a happy customer and sometimes saying no isn't an end in itself.

After you understand the customer and the emotion is controlled, you can proceed to fix the customer. It's useful to know what rules you can break to do this. That's right, break some rules. In aerospace, defense and technology businesses we're rule compliant, not rule averse. In "Find a Problem - Fix a Problem" I noted that statutory law and the Federal Acquisition Regulation cannot be broken, but internal guidelines and policies can be. They might have been put in place for a one-off issue and have continued to burden your company and customers ever since. If these guidelines and policies are irritating one customer, they likely are irritating others. How do you identify them? Engage your Gen X and Millennial colleagues. If they ask you why you're doing something a certain way and you can't give a good answer, check it out. There might be underlying technical, financial or legal reasons, of course, but it just might be that a policy or procedure irritating a customer can be altered or eliminated with positive results both for your customer and your company. Be brave.

Recall that up to 80 percent of customer-client problems involve poor communication. So don't forget to use the 100 percent technique to try and fix a customer before you start talking technical solutions. Say, for example, the customer wants something to work a certain way and it doesn't. Ask them questions to determine just how much ground truth they know about the capabilities and limitations of the product or service. Let's say they know 70 percent of ground truth. Tell them the other 30 percent and let them absorb it and get to 100 percent. You can then help them navigate a work-around. Or you might find that the problem has evaporated. In all cases, you want to be *Jerry Maguire* to your customers: "Help me help you."

Difficult Quotes

"If you want plenty of experience in dealing with difficult people, then have kids."
--Bo Bennett

"Learning to trust is one of life's most difficult tasks."
--Isaac Watts

"When the sun is shining I can do anything; no mountain is too high, no trouble too difficult to overcome."
--Wilma Rudolph

"Do the difficult things while they are easy and do the great things while they are small."
--Lao Tzu

"The Difficult is that which can be done immediately; the Impossible that which takes a little longer."
--George Santayana

"It's difficult to think anything but pleasant thoughts while eating a homegrown tomato."
--Lewis Grizzard

"It is always wise to look ahead, but difficult to look further than you can see."
--Winston Churchill

"The most difficult thing is the decision to act, the rest is merely tenacity."
--Amelia Earhart

"The easiest thing to be in the world is you. The most difficult thing to be is what other people want you to be."
--Leo Buscaglia

"Painting is easy when you don't know how, but very difficult when you do."
--Edgar Degas

"Each success only buys an admission ticket to a more difficult problem."
--Henry A. Kissinger

"Things are not difficult to make; what is difficult is putting ourselves in the state of mind to make them."
--Constantin Brancusi

"Because of their size, parents may be difficult to discipline properly."
--P. J. O'Rourke

"It is a common experience that a problem difficult at night is resolved in the morning after the committee of sleep has worked on it."
--John Steinbeck

"It is difficult to free fools from the chains they revere."
--Voltaire

"It is difficult to get a man to understand something when his salary depends upon his not understanding it."
--Upton Sinclair

"When it becomes more difficult to suffer than to change... you will change."
--Robert Anthony

"There is nothing more difficult to take in hand, more perilous to conduct, or more uncertain in its success, than to take the lead in the introduction of a new order of things."
--Niccolo Machiavelli

"The wireless telegraph is not difficult to understand. The ordinary telegraph is like a very long cat. You pull the tail in New York, and it meows in Los Angeles. The wireless is the same, only without the cat."
--Albert Einstein

"I had been told that the training procedure with cats was difficult. It's not. Mine had me trained in two days."
-- Bill Dana

"Because a thing seems difficult for you, do not think it impossible for anyone to accomplish."
--Marcus Aurelius

"Be thankful for problems. If they were less difficult, someone with less ability might have your job."
--James A. Lovell

"No question is so difficult to answer as that to which the answer is obvious."
--George Bernard Shaw

"When you don't come from struggle, gaining appreciation is a quality that's difficult to come by."
--Shania Twain

"What do we live for, if not to make life less difficult for each other?"
--George Eliot

"There seems to be some perverse human characteristic that likes to make easy things difficult."
--Warren Buffett

"It is far more difficult to be simple than to be complicated."
--John Ruskin

"A chair is a very difficult object. A skyscraper is almost easier. That is why Chippendale is famous."
--Ludwig Mies van der Rohe

"Becoming a leader is synonymous with becoming yourself. It is precisely that simple, and it is also that difficult."
--Warren G. Bennis

"Doing easily what others find difficult is talent; doing what is impossible for talent is genius."
--Henri Frederic Amiel

"It is seldom very hard to do one's duty when one knows what it is, but it is often exceedingly difficult to find this out."
--Samuel Butler

Handshakes

The handshake is extremely important to customer engagement, yet we rarely think much about it. A handshake is natural. Familiar. Everyone does it. What's the big deal? The big deal is that if you're meeting someone important for the first time, a good handshake can get you off on the right foot (so to speak). But a bad handshake can leave a negative impression that's hard to overcome. It's the one time that it's not only permitted, but expected, that we physically touch one another. And as we do, we also judge one another. Too strong, too limp, too lingering can be too bad for your encounter.

Handshakes communicate. For example, there are handshakes that communicate dominance, such as a grasp with the palm slightly down. So if you want to let your key decision makers feel they are in charge, you can angle your palm slightly up. The double-handed shake strives for dominance through intimacy. It can be useful to counter a palm-down power handshake, if it is being forced against your will, but be wary of using this the first time on someone unless you're a politician.

What's a good handshake? The American Psychological Association studied the features that characterize a good handshake and came up with: completeness of grip, vigor, strength, eye contact and duration. The best handshake in almost all situations, according to *The Definitive Book of Body Language* by Barbara and Allan Pease, is a vertical grasp showing respect and equality. Make eye contact, repeat the other person's name and match their grip for firmness. Find balance and make it memorable.

My good friend and protocol expert Kelly Harris teaches these points about handshakes:

- The right hand always should be free (no food or beverage)
- The left hand should hold only one item
- Shoulder-to-shoulder stance
- Extend your hand with the thumb up and fingers out
- Don't extend your hand with the thumb down and fingers curled
- Web-to-web
- Shake from the elbow, not the wrist or shoulder
- Two smooth pumps (the American style)

Kelly also notes that customs differ in other countries. But the American style handshake, perhaps toned down a bit, is accepted worldwide. In some places, neglecting to shake someone's hand is a rude rejection. In all cases, remove your gloves before shaking hands and never shake hands with the other hand in your pocket. In Europe, a woman initiates a handshake. In much of Asia, some women nod slightly, but do not shake hands with men.

For more helpful global handshaking tips, pick up a copy of *Kiss, Bow or Shake Hands*, a good resource for country-by-country handshaking and other tips for business practices and negotiating strategies. If you want to know more about uniformed military customs and courtesies, read the excellent reference *Service Etiquette* by my friends Cherlynn Conetsco and Anna Hart (Naval Institute Press). Cherlynn trained several thousand defense attachés being sent to embassies around the world and Anna created training programs in protocol and etiquette for U.S. Naval Academy Midshipmen.

Handshake Quotes

"Only truthful hands write true poems. I cannot see any basic difference between a handshake and a poem."
--Paul Celan

"Sometimes you don't have to try at all. Sometimes you can hear the wind blow in a handshake."
--Ani DiFranco

"Contracts were made to be broken, honey, but a handshake is the law of God."
--J.R. Ewing to the daughter of an oil business associate

"I got a good handshake. A lot of executives tell me I have the best handshake in Hollywood."
--Marlee Matlin

"There is not a soul who does not have to beg alms of another, either a smile, a handshake, or a fond eye."
--Lord Acton

"An endless mile, a bus wheel turning. A friend to share the lonesome times. A handshake and a sip of wine."
--Charlie Daniels

"I can feel the twinkle of his eye in his handshake."
--Helen Keller

"That is very important. The weak, horrible, wet fish handshake is a problem. That gives a lot away."
--Diana Mather

"A person's word and a man's handshake ought to mean something."
--Joe Glenn

"More history is made by secret handshakes than by battles, bills, and proclamations."
-- John Barth

"Kind words, kind looks, kind acts and warm handshakes, these are means of grace when men in trouble are fighting their unseen battles"
--John Hall

"I like to get within handshaking distance of the crowd. If it happens, they know it, we know it, and that's what we all came here for."
--Levon Helm

"The biggest danger for a politician is to shake hands with a man who is physically stronger, has been drinking and is voting for the other guy."
--William Proxmire

"You cannot shake hands with a clenched fist."
--Indira Gandhi

"A human being's first responsibility is to shake hands with himself."
--Henry Winkler

"I forgot to shake hands and be friendly. It was an important lesson about leadership."
--Lee Iacocca

Selling to **Humans**

Because people want to buy from other people, you don't really sell to government and other businesses. You sell to other human beings. If you're employed in business development and haven't read any of Daniel H. Pink's work, you might want to consider his book, *To Sell is Human: The Surprising Truth about Moving Others.* Dan authored two New York Times best-sellers, *Drive* and *A Whole New Mind,* and then turned his attention to the subject of selling. Since he lives in Washington, D.C., I was fortunate to be at an event where he spoke about why he focused on selling. Pink admits he's old school and cuts out newspaper and magazine articles and stuffs them in folders with nominal titles. Periodically he reviews the folders to see if any are ripe for writing. Some folders he throws away. Others he continues to work on. Over time he became more and more intrigued by the human aspects of selling and that folder grew large. Much to his own surprise, he wrote a book about it.

To engage the reader, Pink begins his book with numbers: one in nine of us earn a living by trying to get others to make a purchase. The other eight in nine are engaged in "non-sales selling." This is persuading, convincing and influencing other people to give up something they have in exchange for something we have. Pink cites statistics that we devote up to 40 percent of our daily activities to moving others toward decisions in our favor. But even though selling is basic to what we all do in our lives, we tend to deny it. It seems that few of us want to be associated with "sales."

Pink conducted a survey of what words came into their heads when people thought of "sales" or "selling." There were very few positive responses. The most frequent words were: pushy, yuck, difficult, hard, ugh, annoying, sleazy, and dishonest. You get the picture. The most common type mentioned was "car salesman." These common attitudes support the general mentality of Buyer Beware (Caveat Emptor), but Pink argues that this image no longer matches reality. Access to information changes the sales interaction. He uses the example of DARCARS, which upended the auto sales industry by openly sharing data with customers and hiring sales staff not previously trained in the auto sales industry. Greater access to information through the internet has transformed the buyer-seller environment to Seller Beware (Caveat Venditor), empowering consumers and allowing them to avoid or ignore "pushy salespeople."

How do you become a better salesperson in this new environment? Pink says it isn't through the old A-B-C of Always Be Closing, but by his new A-B-C of *Attunement, Buoyancy and Clarity*. Attunement: (1) Increase your power by reducing it, (2) Use your head as much as your heart, and (3) Mimic strategically. On this last point, he says the majority of the general population are ambiverts (neither extrovert nor introvert) and well positioned to connect with customers. For Buoyancy, Pink recommends (1) Interrogative self-talk (Can I make a great pitch?) in the preparatory stage, (2) Positivity ratios (the proper mix between levity and gravity) during the day to keep you going, and (3) Explanatory style (clear-eyed optimism) at the end of the day. Regarding Clarity, the keys are (1) Finding the right problems to solve, (2) Finding your frames (comparisons) and (3) Finding off-ramps (giving people details on how to get to where you want them to go).

Because he's interested in the act of moving people to make decisions, Pink takes on the traditional *elevator pitch*, which he claims has at least six variations that can be successfully employed in the new customer environment: the one-word pitch ("Priceless"), the question pitch ("Are you better off now than you were four years ago?"), the rhyming pitch ("If it doesn't fit you must acquit"), the subject-line pitch ("4 tips to improve your golf swing this afternoon"), the Twitter pitch ("Try this link"), and the Pixar pitch from a story artist's template: "Once upon a time ... Every day ... One day ... Because of that ... And because of that ... Until finally ..."

As sales and theater have much in common, it's important that we learn to *improvise* in the new customer environment. Stable, simple and predictable conditions that favored scripted sales presentations have given way to dynamic, complex and unpredictable conditions that, at certain moments, favor a quick change-up. There are virtues in breaking from the script, but we should take lessons from those who are good at improvisational theater. Improvisation may seem like chaos, but it has an underlying structure. The three essential rules of improv theater worth modeling are: (1) Hear offers, (2) Say "Yes, and" (not "Yes, but") and (3) Make your partner look good.

According to Pink, both sales and non-sales selling are ultimately about *service*. Service is more than smiling; it's about making customers' lives better. Customer service can be improved if you (1) Make it personal (give your cell number) and (2) Make it purposeful (for their benefit). Extending the Servant Leadership concept, Pink recommends we think in terms of Servant Selling, acknowledging what the most successful salespeople have long known. In conclusion, the three main messages I heard Dan Pink emphasize while talking about his book were: (1) "Like it or not, we're all in sales now," (2) "We need to get to a state of co-creation with customers," and (3) "Appealing to a higher purpose is a great sales performance enhancer." Ultimately, if you want to be more successful in selling to human beings, you have to try and be more human yourself.

Human Quotes

"Only two things are infinite, the universe and human stupidity, and I'm not sure about the former."
--Albert Einstein

"The universe is not required to be in perfect harmony with human ambition."
--Carl Sagan

"To err is human; to forgive, divine."
--Alexander Pope

"To err is human, but it feels divine."
--Mae West

"To err is human. To blame someone else is politics."
--Hubert H. Humphrey

"All human actions have one or more of these seven causes: chance, nature, compulsions, habit, reason, passion, desire."
--Aristotle

"The main facts in human life are five: birth, food, sleep, love and death."
--E. M. Forster

"Sometimes our light goes out but is blown into flame by another human being. Each of us owes deepest thanks to those who have rekindled this light."
--Albert Schweitzer

"One looks back with appreciation to the brilliant teachers, but with gratitude to those who touched our human feelings."
--Carl Jung

"I long, as does every human being, to be at home wherever I find myself."
--Maya Angelou

"One of the oldest human needs is having someone to wonder where you are when you don't come home at night."
--Margaret Mead

"A wonderful fact to reflect upon, that every human creature is constituted to be that profound secret and mystery to every other."
--Charles Dickens

"You have a nice personality, but not for a human being."
--Henny Youngman

"No institution can possibly survive if it needs geniuses or supermen to manage it. It must be organized in such a way as to be able to get along

under a leadership composed of average human beings."
--Peter Drucker

"I'm not concerned with your liking or disliking me… All I ask is that you respect me as a human being."
--Jackie Robinson

"The care of human life and happiness, and not their destruction, is the first and only object of good government."
--Thomas Jefferson

"I think this is the most extraordinary collection of talent, of human knowledge, that has ever been gathered at the White House - with the possible exception of when Thomas Jefferson dined alone."
--John F. Kennedy

"The human race has one really effective weapon, and that is laughter."
--Mark Twain

"Laughter is the sun that drives winter from the human face."
--Victor Hugo

"No pessimist ever discovered the secret of the stars, or sailed to an uncharted land, or opened a new doorway for the human spirit."
--Helen Keller

"I do not think there is any thrill that can go through the human heart like that felt by the inventor as he sees some creation of the brain unfolding to success… such emotions make a man forget food, sleep, friends, love, everything."
--Nikola Tesla

"Internet is the most important single development in the history of human communication since the invention of call waiting."
--Dave Barry

"The broader one's understanding of the human experience, the better design we will have."
--Steve Jobs

"The greatest discovery of my generation is that a human being can alter his life by altering his attitudes."
--William James

"There is only one basic human right, the right to do as you damn well please. And with it comes the only basic human duty, the duty to take the consequences."
--P. J. O'Rourke

"There are no constraints on the human mind, no walls around the human spirit, no barriers to our progress except those we ourselves erect."
--Ronald Reagan

"The length of a film should be directly related to the endurance of the human bladder."
--Alfred Hitchcock

"Who sees the human face correctly: the photographer, the mirror, or the painter?"
--Pablo Picasso

"It is a bit embarrassing to have been concerned with the human problem all one's life and find at the end that one has no more to offer by way of advice than 'try to be a little kinder'."
--Aldous Huxley

"Every time I see an adult on a bicycle, I no longer despair for the future of the human race."
--H. G. Wells

I

INTEGRITY - *CUSTOMER* **INTIMACY**

Integrity

A story of business Integrity: "I'm working at the store that my new business partner and I put together and this guy comes in to pick up his order. He hands over a crisp new $50 bill. I give him his change and he leaves. As I'm putting the cash in the drawer I notice there are two new fifties stuck together. Now I have a business integrity issue ... do I tell my partner?"

This is the only business integrity joke I know of, and it's lame at that. Integrity and ethics in business are no laughing matter and difficult to write about. I started and stopped this piece several times and just when I'd convinced myself to put it off until later a good friend dropped by and basically dared me to finish it. It's so easy to get preachy about business integrity, but it cuts to the heart of trusted relationships. It seems like we're continuing to lose people we've trusted: sports stars, military heroes, political figures and business leaders. There doesn't seem to be much trust out there – for anyone.

Integrity is learned behavior. My father was a Fort Worth paint salesman. When I was young I didn't fully appreciate what a good man he was. One Saturday when I was in grade school we drove to Buddies Supermarket to buy lawn chairs. We paid for four chairs and a stack of them was put in the back of our station wagon. When we unloaded the chairs at home we found we'd been given five by mistake. Without hesitating, Dad drove us back to the store to hand over the extra chair to an astonished store manager. I asked Dad why didn't we just keep the extra chair, as it was their mistake after all? He looked at me knowingly, asked me to think about it and said that one day I'd understand.

I've told that story many times over the years and have tried to live up to that lesson. Recently my son recounted that he'd ordered two bar stools for his Washington DC condo, but three came in the box. He realized that three stools looked nicer at his bar than two, so instead of returning the extra stool he called the store, told them what happened and offered to pay for it. Even after confirming that it wasn't a crank call, the salesman was reluctant to do anything. Eventually my son dogged the supervising manager into taking more money. Many times the right thing to do is the hardest. I know, because as my son was telling me his story, full of justifiable pride for having followed the fine example set by his grandfather, I was embarrassed in recalling my own shortcomings. In business can we be as honest as our parents? Can we do the right thing like our children? I hope so.

Business Integrity Quotes

"The most important persuasion tool you have in your entire arsenal is integrity."
--Zig Ziglar

"Relativity applies to physics, not ethics."
--Albert Einstein

"Never let your sense of morals get in the way of doing what's right."
--Isaac Asimov

"My father was very strong. I don't agree with a lot of the ways he brought me up. I don't agree with a lot of his values, but he did have a lot of integrity, and if he told us not to do something, he didn't do it either."
--Madonna Ciccone

"If everyone were clothed with integrity, if every heart were just, frank, kindly, the other virtues would be well-nigh useless."
--Moliere

"I never questioned the integrity of an umpire. Their eyesight, yes."
--Leo Durocher

"Subtlety may deceive you; integrity never will."
--Oliver Cromwell

"Once you get rid of integrity the rest is a piece of cake."
--J.R Ewing (Larry Hagman)

"Integrity is so perishable in the summer months of success."
--Vanessa Redgrave

"Integrity without knowledge is weak and useless, and knowledge without integrity is dangerous and dreadful."
--Samuel Johnson

"Integrity is the essence of everything successful."
--R. Buckminster Fuller

"Integrity has no need of rules."
--Albert Camus

"There are seven things that will destroy us: Wealth without work; Pleasure without conscience; Knowledge without character; Religion without sacrifice; Politics without principle; Science without humanity; Business without ethics."
--Mahatma Gandhi

"It takes less time to do a thing right than to explain why you did it wrong."
--Henry Wadsworth Longfellow

"Ethics is knowing the difference between what you have a right to do and what is right to do."
--Potter Stewart

"If you don't know where you are going, you'll end up someplace else."
--Yogi Berra

"Real integrity is doing the right thing, knowing that nobody's going to know whether you did it or not."
--Oprah Winfrey

"If it is not right do not do it; if it is not true do not say it."
--Marcus Aurelius

"Be good to your work, your word, and your friend."
--Ralph Waldo Emerson

"I will not deny but that the best apology against false accusers is silence and sufferance, and honest deeds set against dishonest words."
--John Milton

"I am not bound to win, but I am bound to be true. I am not bound to succeed, but I am bound to live up to what light I have."
--Abraham Lincoln

"A quiet conscience makes one strong!"
--Anne Frank

"In a room where people unanimously maintain a conspiracy of silence, one word of truth sounds like a pistol shot."
--Czesław Miłosz

"Even the most rational approach to ethics is defenseless if there isn't the will to do what is right."
--Alexander Solzhenitsyn

"Somebody once said that in looking for people to hire, you look for three qualities: integrity, intelligence, and energy. And if you don't have the first, the other two will kill you."
--Warren Buffett

"Our deeds determine us, as much as we determine our deeds."
--George Eliot

"I hope I shall always possess firmness and virtue enough to maintain what I consider the most enviable of all titles, the character of an honest man."
--George Washington

"I am different from Washington; I have a higher, grander standard of principle. Washington could not lie. I can lie, but I won't."
--Mark Twain

"The way to a good reputation is to endeavor to be what you desire to appear."
--Socrates

"To be persuasive we must be believable; to be believable we must be credible; credible we must be truthful."
--Edward R. Murrow

"A liar needs a good memory."
--Quintillian

"Lead your life so you wouldn't be ashamed to sell the family parrot to the town gossip."
--Will Rogers

Customer Intimacy

Customer intimacy. Such a tricky subject. Years ago I was invited to speak on customer relations and the host introduced me with "Here's Dave Potts, who has a real passion for customer intimacy." Creepy. During my initial survey of customer relations and sales literature years ago I kept coming across the term customer intimacy. I began to notice that it frequently came up in senior executive discussions *after* there was a problem with a program and even then there was no general agreement of what the term meant or what to do about it other than to punish or swap out people.

It's true that sometimes the chemistry between government customers and aerospace, defense and technology contractors can be bad. But before you make difficult and costly personnel changes, consider that poor business relationships could be caused by a lack of comprehending and appreciating the human aspects of the customer-contractor relationship. People want to buy from other people, especially if the products and services are complex and highly technical. So it might be worthwhile to explore what customer intimacy is and where it fits in the customer relations domain.

I believe that customer intimacy is the sum of what we do every day with our customers. There is frequency of contact. Do you know if your customer wants lots of "touches" or wants you not to bother them until they need you? There is depth of knowledge. When a customer needs information, do you have it? There is range of coordination. Are you connected to everyone on your side and the customer's side so you can forecast issues before they occur? And there is timeliness of response. Are you prompt in answering customer queries?

The "what" you do in customer intimacy is vital, but because your customers want human connectivity, I believe the "how" can be even more important. Are you open to communication? Do you listen and try to understand? Do you project respect and trust (even to difficult customers)? Finally, and most important, are you easy to work with (See the Z section)? Together with *what*, don't forget the *how*.

Where does customer intimacy fit in the customer relations domain? There are both internal and external aspects of customer relations. The internal aspects of customer relations, such as leadership emphasis on fostering corporate-wide customer-centric culture. There are customer relations skills training courses that wise companies invest in to support that customer-centric culture. And there are business processes that should be ever mindful of the importance of the customer in all of your internal decision-making. These are the main internal aspects of customer relations.

For external aspects, many senior executives mistakenly equate satisfaction and performance metrics with customer intimacy. These certainly are important and necessary in gauging how well or bad you've been doing. But because most industry metrics are lagging indicators, they don't help too much with what's going on in real time with your customers. Some businesses develop customer intimacy indexes with the hope of applying metrics to evaluate business relationships. But usually they're filled out internally and don't include customer input. Customer surveys, applied appropriately, can help in some cases and customers themselves can produce their own ratings (such as Defense Department CPARS). But metrics are not intimate.

Some senior leaders equate branding with customer intimacy, but that misses the mark. As an aspect of customer relations, branding is the external face of your enterprise to the customer base and public at large. Branding can help put positive images and feelings of your products, services, and people into the minds of customers, but it's one-way communication. Customer intimacy involves two-way communication. I am convinced that customer intimacy can be the strongest pillar in supporting and sustaining great customer relationships - more important than innovative solutions and program performance. And it's the one aspect, either internal or external, you have the most control over.

My colleague Dr. Tom Barrett often says that a customer wants to know three things: Do you care? Do I trust you? Do you have anything to say? It's interesting that the first two questions deal with feeling before the third deals with facts. When you're selling, you first sell yourself, then your company, and finally your solution. What do most business developers usually lead with? The solution. In customer intimacy, your customer has to buy you and trust you before anything else can happen.

Does your customer believe you care? If you want to assess intimacy with your government customers, here's my C.A.R.E. model, based on best customer behaviors. Intimate customers are:

- <u>Communicative</u>
  - Tell you first – not newspaper reporters or their lawyers
  - Ask you questions – and listen to your answers
  - Drop hints – wanting you to win
  - Know you – even what you do on the weekends
- <u>Active</u>
  - Do as much as possible – within the bounds of the contracts
  - Are open to new ideas – "What if we …?"
  - Support contracts – and don't begin negotiating after signing
  - Think long term – avoiding panic at minor setbacks
- <u>Responsive</u>
  - Return your calls – and if they don't, what do you do?
  - Are confident and competent – capable of making timely decisions
  - Accommodate schedules – offering to meet when you're in town
  - Are willing to visit you – and come to your demonstrations
- <u>Engaged</u>
  - Use "we" in discussions – and bring that term up first
  - Ask you for advice – "What are we going to do?"
  - Extend your business – Good customers avoid recompetes
  - Identify with your success – (See B.I.R.G. in Emotions)

If your customer CAREs, you have great customer intimacy. And if there's any one aerospace and defense leader I can think of who had great customer intimacy, it's Norm Augustine. He worked in several defense companies, served in government and taught at universities. Norm had an uncanny ability to disarm customers with humor and then win them over with logic and facts. Here are some quotes that give you an idea of why he connected so well with business leaders, government decision makers and university professors.

Norman R. Augustine Quotes

"Bulls do not win bull fights. People do."

"Decreased business base increases overhead. So does increased business base."

"The process of competitively selecting contractors to perform work is based on a system of rewards and penalties, all distributed randomly."

"Rank does not intimidate hardware. Neither does the lack of rank."

"The best way to make a silk purse from a sow's ear is to begin with a silk sow. The same is true of money."

"Although most products will soon be too costly to purchase, there will be a thriving market in the sale of books on how to fix them."

"If you can afford to advertise, you don't need to."

"One of the most feared expressions in modern times is, 'The computer is down'."

"In my view the organization has been far more successful than I dreamed it would be. But my view is also that it's still an unproved experiment."

"Simply stated, it is sagacious to eschew obfuscation."

"Regulations grow at the same rate as weeds."

"Two-thirds of the Earth's surface is covered with water. The other third is covered with auditors from headquarters."

"If the Earth could be made to rotate twice as fast, managers would get twice as much done. If the Earth could be made to rotate twenty times as fast, everyone else would get twice as much done since all the managers would fly off."

"The weaker the data available upon which to base one's conclusion, the greater the precision which should be quoted in order to give the data authenticity."

"Software is like entropy. It is difficult to grasp, weighs nothing, and obeys the second law of thermodynamics; i.e. it always increases."

"There are no lazy veteran lion hunters."

"If sufficient number of management layers are superimposed on top of each other, it can be assured that disaster is not left to chance."

"In the year 2054, the entire defense budget will purchase just one aircraft. This aircraft will have to be shared by the Air Force and Navy 3-1/2 days each per week except for leap year, when it will be made available to the Marines for the extra day."

"All too many consultants, when asked, 'What is 2 and 2?' respond, 'What do you have in mind'?"

"By the time the people asking the questions are ready for the answers, the people doing the work have lost track of the questions."

"There are many highly successful businesses in the United States. There are also many highly paid executives. The policy is not to intermingle the two."

"If today were half as good as tomorrow is supposed to be, it would probably be twice as good as yesterday was."

"Any task can be completed in only one-third more time than is currently estimated."

"Fools rush in where incumbents fear to tread."

"A billion saved is a billion earned."

"One-tenth of the participants produce over one-third of the output. Increasing the number of participants merely reduces the average output."

"The last 10 percent of performance generates one-third of the cost and two-thirds of the problems."

"A revised schedule is to business what a new season is to an athlete or a new canvas to an artist."

"Most projects start out slowly - and then sort of taper off."

"One should expect that the expected can be prevented, but the unexpected should have been expected."

"Simple systems are not feasible because they require infinite testing."

"Hardware works best when it matters the least."

"It's easy to get a loan unless you need it."

"It is very expensive to achieve high unreliability. It is not uncommon to increase the cost of an item by a factor of ten for each factor of ten degradations accomplished."

"Aircraft flight in the 21st century will always be in a westerly direction, preferably supersonic, crossing time zones to provide the additional hours needed to fix the broken electronics."

"The average regulation has a life span one-fifth as long as a chimpanzee's and one-tenth as long as a human's - but four times as long as the official's who created it."

J

JOLT YOUR POWERPOINTS AND PRESENTATIONS

Jolt Your PowerPoints and Presentations

I know, I know. Why put PowerPoints and Presentations in the J section? Simple. The P section was too crowded and I couldn't think of any customer relations and sales subject I wanted to discuss that began with J. And that, actually, is to the point. You often have to make do with what you have to work with, such as PowerPoint presentations. Don't misinterpret the message. I love PowerPoint. I'm of the generation that can remember grease pencils on transparencies, then clunky Harvard Graphics and then all the other competitors to Microsoft PowerPoint. There was an electronic Tower of Babel, which thankfully is now history.

The problem with slides and charts today is not technical; it's a lack of understanding of when and how to use PowerPoint presentations. This is similar to a business that gets all wrapped up in developing a product before it understands what it's to be used for. PowerPoint as a medium is so overwhelming in its capabilities and capacities that it's now used in place of white papers, decision briefs, marketing presentations, informational lectures and, too often, group discussions. One colleague told me that the most effective team offsite she had ever attended had no PowerPoints. The team lead banned them. Everyone had to stand up and deliver their reports with no notes, but could use visual aids and draw on flip charts. It was a back-to-basics meeting that was both productive and fun. In contrast, another colleague said his vice president decided to forego all slides in a key leaders meeting and it was a disaster because no one could remember what he said. To me, these two examples illustrate that PowerPoint can be very useful, but shouldn't be used in all cases. It's all about getting the right messages across in the right way to the right audience.

PowerPoints

The first question that should be asked is about purpose. What will this PowerPoint be used for? If it's a white paper it should show how relevant the offered technology is to the viewer. If it's a decision brief it should have courses of action and a recommendation. If it's a marketing presentation it should capture the viewer's attention and interest. If it's a personnel training program it should be informative and actionable. If it's a technical program review it should be constructed with lots of clear data charts. Where you are in the life of a program and what outcome you want to achieve determine the kind of PowerPoint presentation you should construct, or whether you should even use a PowerPoint at all.

Years ago George Lucas of Star Wars fame was on a late night talk show and was asked how he made great movies. Lucas replied that he put together a hot opening and a hot closing and tried not to screw up the middle. That's great advice for a PowerPoint presentation. Unfortunately, many aerospace, defense and technology business developers aren't comfortable with marketing. That's why so many of their PowerPoints fall flat and fail to have impact with customers. A good customer-facing PowerPoint isn't a corporate overview or branding advertisement – it's a marketing tool to convince specific customers that their lives will be better with your products and services.

Fill the gap between the opening and closing of your presentation with three main points. This is, after you grab attention with the hot opening and before you set up a hot closing and call to action. Why limit yourself to three main points? Because people normally can only remember three things at a time. That's why so many "sticky" ads are constructed that way. This principle isn't rocket science or anything new, but we tend to forget it. In the Air Force, I was trained in staff work that decision packages to flag officers should be limited to one page and three main points (with attachments to support the recommendations). Flag officers normally have attention deficit issues based on their busy schedules. One page and three main points make it easy for them to understand problems and solutions.

George Lucas also said, "A special effect without a story is a pretty boring thing." We've all been victimized by pointless, nauseating motion in PowerPoint presentations. But that doesn't mean they should be constructed like stone tablets. A mixture of words, pictures and videos can engage all learning channels of your customers and, properly prepared, can make your presentation meaningful and memorable. I'm particularly taken with imbedded videos. Videos are the next best thing to a demonstration and draw in the customer's attention by showing what your product or service actually does. This technique is particularly good for Gen X and Millennial customers.

A great PowerPoint slide stands on its own. It's a self-evident story, in key words or solution pictures or in both. A great PowerPoint slide can be briefed in as little as one minute or as many as ten. If you'd rather spend a customer meeting in discussion, you can leave a great PowerPoint slide behind as a summary of the visit and something to be shared. If you're a master marketer, a great PowerPoint slide is of such interest to customers that they'll incorporate it into their own presentations up the decision chain.

Finally, I would share the caution of Lockheed's Kelly Johnson in a 1963 memo to staff forbidding "idiot charts," defined as:

- "One that states by written sentence a subject which is perfectly clear to the audience and which should be to the presenter. It is generally designed to allow the presenter to spend four or five minutes on things which have nothing to do with the chart."
- "A systems diagram which looks very much like all other systems diagrams, in that it consists of square boxes starting with an input from some source and ending up with an output. A great deal of time can be wasted with such artwork and very seldom does it make any point..."

If you are a PowerPoint power enthusiast, there are many good techniques available for download from the web. If you remain a PowerPoint skeptic, here are some quotes that will reinforce your mindset. Like PowerPoint, something for everyone.

PowerPoint Quotes

"You can't speak with the U.S. military without knowing PowerPoint."
--Margaret Hayes, National Defense University

"Power corrupts and PowerPoint corrupts absolutely."
--Vint Cerf, Internet pioneer

"If your words or images are not on point, making them dance in color won't make them relevant."
--Edward Tufte, Professor Emeritus, Yale University

[Michael]: "We'll ask PowerPoint." [Oscar]: "Michael, this is a presentation tool." [Michael]: "You're a presentation tool!"
--*The Office*

"Despite the level of cadet complaints about the 'Death by PowerPoint' phenomena, I have found that the cadets are quite willing to inflict this upon their colleagues."
--U.S. Military Academy Faculty

"PowerPoint is a triumph of process over product. Knowing what you are doing is more important than getting the right answer."
--Tom Lehrer, mathematician, singer-songwriter

"The genius of it is that it was designed for any idiot to use. I learned it in a few hours."
--David Byrne, *Talking Heads*

"My belief is that PowerPoint doesn't kill meetings. People kill meetings."
--Peter Norvig, Google, Inc.

"The idea behind most of these briefings is for us to sit through 100 slides with our eyes glazed over, and then to do what all military organizations hope for ... to surrender to an overwhelming mass."
--Richard Danzig, former Secretary of the Navy

"The 'PowerPoint Syndrome' is a well known disease, clearly diagnosed not only by brilliant cartoonists such as Scott Adams, but also in a variety of analyses of corporate efficiency and communication. It's called disinfotainment."
--Giancarlo Livraghi, Italian scientist

"I generally believe that PowerPoint is the spawn of Satan. It breeds passivity in the students and it disconnects the speaker from the audience."
--Professor John Arras, University of Virginia

"It's like a plastic banana…looks good but provides no nutritional value or sustenance."
--Anonymous Lieutenant Colonel in the National Capitol Region

"I must say I started to see more bad plans with good slides approved over good plans with no slides."
--Robert Walsh, President and CEO of EnvisionWare

"While you were making your slides, we would be killing you."
--Russian Officer to C.M. Coglianese during a discussion between U.S. and Russian officers serving in Bosnia as to who would have won if we had ever actually fought in Western Europe.

[Buffy]: "So I can skip the PowerPoint presentation, huh?"
--*Buffy the Vampire Slayer Season Eight*

Presentations

When you're preparing to present, first consider your audience, then your message, then how you will deliver *that* message to *that* audience. Who's the audience and what's in it for them to listen to you? If it's a general audience, say at a trade show, then an overview could be appropriate. If it's a technical audience, you need to impart lots of well-organized and clear data. If it's an executive buyer, then you likely should focus on value. Of course, there are differing types of value. I can remember preparing a senior business executive for presentations to both a minister of defense and a minister of finance on the same day. We were giving the same presentation and same message, but with different value emphasis. Due to jet lag we managed to present economic value to the minister of defense and security value to the minister of economy. Not a good day.

If you haven't thought through what your messages are, no great PowerPoint tricks will help. Many communication experts say you should have three main points. Power Messaging courses by Corporate Visions teach that each of your three main points, or Power Messages in their terminology, should contain three essential elements. First, it should be of value to the customer. It should line up with your value proposition and win themes. Second, it should be easy to defend. It should create an "Aha" moment in the mind of the customer. Third, it should be unique to your company. If it's not unique to you, then you're not differentiating yourself from your competition and muddling the reasons why the customer should pick you.

Your three main points, taken together with an attention-grabbing opening and an emotional call-to-action closing is your overall story. What's your story? I often found it amusing that storyboards in proposal rooms are used to construct business proposals that lack an apparent story. Stories make all the difference, no matter if it's a PowerPoint presentation, a written technical proposal or a speech given in an auditorium. Storytellers captivate our imagination. Data dumpers put us to sleep. Just like systems within systems, there are stories within stories. Each main point should have its own story and those stories should reinforce each other. Stories are as old as campfires and tribal elders. We pay attention when a speaker says, in effect, "Once upon a time…" A well-practiced and well-placed story can move the audience emotionally toward you and the point you want to make.

After you've analyzed audiences and created tailored messages, you can turn attention to yourself and your delivery. If you're reading this book alphabetically, recall Dr. Tom Barrett's ACE model. Authority, Conviction and Enthusiasm can help you master all the other things you need to consider in preparing your presentation. Above all, remember that you are the most important component in delivering your messages (see *You are the Message* by Roger Ailes). But that can be daunting. Stay authentic to keep yourself grounded. Don't try to be something you're not. For example, if you can't tell a joke well, don't tell one.

Authenticity (which comes after Alternatives in the A section) can help you keep a clear head while you must do many things at the same time to communicate effectively. For example, if you play golf you know that if you worry too much about keeping your knees bent you may not roll your wrists correctly (or one of the countless other things necessary to hit that little white ball correctly). But if you're comfortable being yourself and practice your moves, you can keep the ball in the fairway. You may not be as good as Arnold Palmer as a golfer or Steve Jobs as a presenter, but you can deliver a winning presentation if you stay Authentic and be an ACE.

Essentially all presentation experts say you should consider the Visual, Vocal and Verbal: your body language, voice tone and choice of words. Body language is covered in section X, voice tone in section V and words in section W. Although you shouldn't teach negatives, we're all interested in learning from the mistakes of others (see Lessons Learned in section L). Communications coach and author Carmine Gallo has a list of the ten worst habits he suggests avoiding if you want your presentations to "sing."

1. Reading from notes
2. Avoiding eye contact
3. Dressing down
4. Fidgeting, jiggling and swaying
5. Failure to rehearse
6. Standing at attention
7. Reciting bullet points
8. Speaking too long
9. Failing to excite
10. Ending with an inspiration deficit

But all the tips from all the experts don't make it any easier to get up in front of an audience and begin to speak. As a young man I was a terrible speaker. My heart would race, my brain would fog and I would rush over my words and garble my sentences. Yet I really liked being the center of attention. My dear mother used to remind me, "David, the world doesn't revolve around you." I still think she was wrong. What helped me at first was a public speaking course I took as an undergraduate. I found out that speaking skills could be learned and if I knew my material well my fear of speaking would fade.

What helped me next were all the opportunities I had in the Air Force to get up in front of audiences and make mistakes. After hearing one of my disasters that ended with a general officer glaring at me, a mentor remarked, "Potts, I've never seen a captain in that much trouble before." We laughed as I learned. You need to road test your presentation skills. Years later I advised a young man working for me to take a Toastmasters course. He was super-smart, but couldn't speak intelligibly until, like me, he learned from training and experience. Now he runs his own multi-million dollar company.

The third thing that helped me was taking time to analyze how expert speakers work their audiences. These were politicians, industry titans and motivational speakers. I borrowed here and there, tried things out and kept the techniques that worked for me. I'm still not fully satisfied with my public speaking, but I can get an audience's attention, hold it with stories, make three main points, and close with emotion. I hope this gives you encouragement to make your next presentation the one they won't forget.

Presentation Quotes

"There are always three speeches, for every one you actually gave. The one you practiced, the one you gave, and the one you wish you gave."
--Dale Carnegie

"You can speak well if your tongue can deliver the message of your heart."
--John Ford

"The audience only pays attention as long as you know where you are going."
--Philip Crosby

"Let thy speech be better than silence, or be silent."
--Dionysius of Halicarnassus

"No one ever complains about a speech being too short!"
--Ira Hayes

"According to most studies, people's number one fear is public speaking. Number two is death. Death is number two. Does that sound right? This means to the average person, if you go to a funeral, you're better off in the casket than doing the eulogy."
--Jerry Seinfeld

"They expect a professional presentation, so they expect to see a 'professional.' Dress appropriately for the occasion, but don't be one of the crowd."
--Wess Roberts

"The human brain starts working the moment you are born and never stops until you stand up to speak in public."
--George Jessel

"It usually takes more than three weeks to prepare a good impromptu speech."
--Mark Twain

"A theme is a memory aid; it helps you through the presentation just as it also provides the thread of continuity for your audience."
--Dave Carey

"Today's public figures can no longer write their own speeches or books, and there is some evidence that they can't read them either."
--Gore Vidal

"What this country needs is more free speech worth listening to."
--Hansell B. Duckett

"There are two things that are more difficult than making an after-dinner speech: climbing a wall which is leaning toward you and kissing a girl who is leaning away from you."
--Winston Churchill

"The duty of a toastmaster is to be so dull that the succeeding speakers will appear brilliant by contrast."
--Clarence B. Kelland

"A good orator is pointed and impassioned."
--Cicero

"Make sure you have finished speaking before your audience has finished listening."
--Dorothy Sarnoff

"The most precious things in speech are the pauses."
--Sir Ralph Richardson

"Speak when you are angry—and you will make the best speech you'll ever regret."
--Laurence J. Peter

"Every speaker has a mouth; An arrangement rather neat. Sometimes it's filled with wisdom. Sometimes it's filled with feet."
--Robert Orben

"All of us are born with a set of instinctive fears - of falling, of the dark, of lobsters, of falling on lobsters in the dark, or speaking before a Rotary Club, and of the words 'Some Assembly Required'."
--Dave Barry

"Public speaking is the art of diluting a two-minute idea with a two-hour vocabulary."
--John Fitzgerald Kennedy

"Before I speak, I have something important to say."
--Groucho Marx

"Ask yourself, 'If I had only sixty seconds on the stage, what would I absolutely have to say to get my message across'."
--Jeff Dewar

"To talk well and eloquently is a very great art, but that an equally great one is to know the right moment to stop."
--Wolfgang Amadeus Mozart

"Grasp the subject, the words will follow."
--Cato the Elder

"Once you get people laughing, they're listening and you can tell them almost anything."
--Herbert Gardner

"Say not always what you know, but always know what you say."
--Claudius

"Words have incredible power. They can make people's hearts soar, or they can make people's hearts sore."
--Dr. Mardy Grothe

"Speech is power: speech is to persuade, to convert, to compel."
--Ralph Waldo Emerson

"It's not how strongly you feel about your topic, it's how strongly they feel about your topic after you speak."
--Tim Salladay

"The power of sound has always been greater than the power of sense."
--Joseph Conrad

"Many attempts to communicate are nullified by saying too much."
--Robert Greenleaf

"Words are, of course, the most powerful drug used by mankind."
--Rudyard Kipling

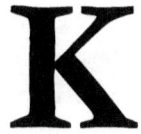

KEEP IT SHORT - KILLER ARGUMENT

Keep It Short

Sometime back a friend sent me a link to a blog by Jason Baer: "Get Shorty – The Elevator Pitch is Dead". Calling to mind Moore's Law about computing power increasing unabated, Baer named a new law: The Law of Boundless Brevity. The idea is that over time, all communication becomes steadily truncated. He suggests that in our high-speed environment we no longer have the ability and opportunity to ponder lengthy phrases. An elevator ride now seems too brief, so the elevator pitch needs to be abandoned in favor of something shorter. Baer suggests a pitch of no more than 120 characters (allowing a re-tweet).

About the same time I saw that blog I read an article in the Washington Post on a problem the Washington Area Metro Transit Authority was having with their alert system. Attempting to keep riders more informed, Metro set up a twitter account to broadcast alerts about subway and bus problems. Great idea to get the word out, right? The problem that erupted was not in the technology of the system or lack of customer demand, but with their own well-meaning Metro employees who were sending out the alerts. They just couldn't keep their messages within the 140-character Twitter limit, so 60 percent of all alert messages were being truncated. An example message was: "No line: There is no Blue line train service between Rosslyn & King Street. Shuttle bus service is established. Customers are encouraged to…" This particular message gave rise to an informal web-based contest to complete the sentence. Some of the entries were: "ford the Potomac at their own risk," "go to the nearest bar and drink away the stress," and "walk, as that will probably be faster." In truth, Washington Metro has since fixed that problem. It's really a great way to get around DC.

I also came across new research that showed the average American can only listen for 28 seconds before getting distracted. Like everyone, you struggle to get your complicated business messages across to frazzled customers (and bosses) in the most effective way. Do you remember back in school when the teachers had you create a poem in Japanese haiku? Haiku is an ancient poetic form consisting of 17 (or fewer) syllables in three metrical phrases of five, seven and five respectively. While traditional haiku is reflective of nature and was printed on a single line with a picture, its modern English equivalent consists of three lines reflective of the urban environment. Here's a 60-character haiku example in proper form by the humorist Mad Kane:

Suffocating spam
Pours into my computer,
Drowning out meaning.

Maybe the elevator pitch really is dead. Perhaps a better pitch could be delivered to distracted customers by thinking in terms of a Twitter haiku. Here's my 83-character Twitter haiku reminder about crisp customer communication:

Don't forget to speak
Using your best friend's voice mode
And with three main points.

What's your short message? While you're thinking about it, here's a brief list of brevity quotes.

Brevity Quotes

"The present letter is a very long one, simply because I had no leisure to make it shorter."
--Blaise Pascal

"It is my ambition to say in ten sentences what other men say in whole books - what other men do not say in whole books."
--Friedrich Wilhelm Nietzsche

"Therefore, since brevity is the soul of wit, and tediousness the limbs and outward flourishes, I will be brief."
--William Shakespeare

"The most valuable of all talents is that of never using two words when one will do."
--Thomas Jefferson

"If you bring that sentence in for a fitting, I can have it shortened by Wednesday."
--Hawkeye in M*A*S*H

"If any man will draw up his case, and put his name at the foot of the first page, I will give him an immediate reply. Where he compels me to turn over the sheet, he must wait my leisure."
--Lord Sandwich

"Let thy speech be short, comprehending much in few words."
--The Bible, Ecclesiastes

"Spartans, stoics, heroes, saints and gods use short and positive speech."
--Ralph Waldo Emerson

"Brevity is the best recommendation of speech, whether in a senator or an orator."
--Cicero

"Talk low, talk slow, and don't say too much."
--John Wayne

"The fewer the words, the better the prayer."
--Martin Luther

"Be sincere; Be brief; Be seated."
-- Franklin Delano Roosevelt

"A short saying often contains much wisdom."
--Sophocles

"If I am to speak ten minutes, I need a week for preparation; if fifteen minutes, three days; if half an hour, two days; if an hour, I am ready now."
--Woodrow Wilson

"Whatever advice you give, be brief."
--Horace

Killer Argument

The one great "killer" argument in a customer sales presentation is, "We've Done That." More than anything else, this simple statement reduces risk in the buyer's mind. And coupled with tailored ROI facts, it can set up a compelling call to action and a customer decision to partner with your company.

Recall that the customer first buys you, then your company and finally your solution. Most business developers usually lead with their brilliant solution, but sometimes the customer isn't ready to buy it yet. The critical first step is to connect with customers on a human level, but don't overlook that you have to sell your company too. It helps if your calling card has the logo of an established brand, but even then your customer may only know your company from a few of your most well-established programs and may be unfamiliar with breadth and depth of the firm. Don't bore customers with organization charts and rah-rah statements. Focus their attention on what your company does.

I believe there is a tendency to understate the incredible things that are being done in the aerospace, defense, and technology sectors. Science fiction is becoming fact:

- Do you want to send a probe to another planet? We've Done That.
- Do you want a satellite that can give your exact position? We've Done That.
- Do you want a platform that can fly through the atmosphere at Mach 20? We've Done That.
- Do you want an airplane that's invisible to radar? We've Done That.
- Do you want a missile that can hit a missile? We've Done That.
- Do you want a warship that's so fast it can outrun a speedboat? We've Done That.
- Do you want a combat vehicle that survives bomb blasts? We've Done That.
- Do you want your national archives digitized? We've Done That.

- Do you want all your computer systems to work seamlessly together? We've Done That.

Of course, there are some instances when nobody's done that. That's when strategic business partners can join in and past performance comes into play. You may be good at pulling teams together to meet specific customer requirements, but customer-felt trust is more often generated by evidence of sustained excellence. Because some companies tend to be forward-looking in line with their technical offerings, they're reluctant to talk about their many years of sustained excellence. It's good to take a moment now and then to look back and remind yourself of the astounding contributions American aerospace, defense and technology companies have made to national security and the wellbeing of the world. And by doing that, you can remind your customers that you were there yesterday when they needed you, and you'll be there tomorrow when they'll need you.

There's probably no one person who contributed more to allowing Lockheed to say "We've Done That" than Kelly Johnson, the legendary leader of the Skunk Works. You can read about Kelly in his autobiography, *Kelly: More Than My Share of It All*, and in *Skunk Works* by his protégé Ben Rich, which is one of the more informative and entertaining books about the aircraft industry. Kelly didn't really talk about his work a lot, but he did bequeath us his (in)famous Rules of Management, which remain timeless advice for all government programs and managers.

Kelly Johnson's 14 Rules of Management

1. The Skunk Works manager must be delegated practically complete control of his program in all aspects. He should report to a division president or higher.
2. Strong but small project offices must be provided both by the military and industry.
3. The number of people having any connection with the project must be restricted in an almost vicious manner. Use a small number of good people (10 to 25 percent compared to the so-called normal systems).
4. A very simple drawing and drawing release system with great flexibility for making changes must be provided.

5. There must be a minimum number of reports required, but important work must be recorded thoroughly.
6. There must be a monthly cost review covering not only what has been spent and committed but also projected costs to the conclusion of the program. Don't have the books ninety days late and don't surprise the customer with sudden overruns.
7. The contractor must be delegated and must assume more than normal responsibility to get good vendor bids for subcontract on the project. Commercial bid procedures are very often better than military ones.
8. The inspection system as currently used by the Skunk Works, which has been approved by both the Air Force and Navy, meets the intent of existing military requirements and should be used on new projects. Push more basic inspection responsibility back to subcontractors and vendors. Don't duplicate so much inspection.
9. The contractor must be delegated the authority to test his final product in flight. He can and must test it in the initial stages. If he doesn't, he rapidly loses his competency to design other vehicles.
10. The specifications applying to the hardware must be agreed to well in advance of contracting. The Skunk Works practice of having a specification section stating clearly which important military specification items will not knowingly be complied with and reasons therefore is highly recommended.
11. Funding a program must be timely so that the contractor doesn't have to keep running to the bank to support government projects.
12. There must be mutual trust between the military project organization and the contractor with very close cooperation and liaison on a day-to-day basis. This cuts down misunderstanding and correspondence to an absolute minimum.

13. Access by outsiders to the project and its personnel must be strictly controlled by appropriate security measures.
14. Because only a few people will be used in engineering and most other areas, ways must be provided to reward good performance by pay not based on the number of personnel supervised.

LESSONS LEARNED - LISTENING

Lessons Learned

Every February 2nd on Groundhog Day, Punxsutawney Phil comes out and we get a "furcast." If it's cloudy and there's no shadow, then spring will come soon. If it's sunny and Phil sees his shadow, then we'll have six more weeks of winter. Even with this guidance it's hard to remember what happens every year with Phil, which brings to mind the memorable 1993 "Groundhog Day" film starring Bill Murray and Andie MacDowell, where Murray played Phil Connors, a weatherman who finds himself living the same day over and over again. In case you need a reminder, here are some quoted excerpts:

> Phil Connors: *Excuse me, where is everybody going?*
> Fan on Street: *To Gobbler's Knob. It's Groundhog Day.*
> Phil Connors: *It's still just once a year, isn't it?*

> Phil: *Do you know what today is?*
> Rita: *No, what?*
> Phil: *Today is tomorrow. It happened.*

> Phil: *Do you ever have déjà vu, Mrs. Lancaster?*
> Mrs. Lancaster: *I don't think so, but I could check with the kitchen.*

> Phil: *What would you do if you were stuck in one place and every day was exactly the same, and nothing that you did mattered?*
> Ralph: *That about sums it up for me.*

> Rita: *Do you ever have déjà vu?*
> Phil: *Didn't you just ask me that?*

Phil: *This is one time where television really fails to capture the true excitement of a large squirrel predicting the weather.*

Rita: *This day was perfect. You couldn't have planned a day like this.*
Phil: *Well, you can. It just takes an awful lot of work.*

As you recall, Phil eventually sorts things out by putting knowledge gained from (spectacular) failures to good use. Until he implemented lessons learned, however, he just kept experiencing the same thing again and again. In your business you may have seen something like this. People tend to do the same things well when they win and make the same mistakes when they lose. We could call the latter *lessons experienced*, rather than lessons learned. It's Groundhog Day – over and over and over.

Does your business gather lessons learned and share them? If it does, you've probably noticed trends. When you win you figured out value to the customer and then crafted your solution and messaging based on that perceived value. You developed clear strategies and capture plans and followed disciplined processes developed by seasoned business developers. You gained company-wide support and sufficient resources for your effort and put the right team in place to execute the plan. Everyone on the team knew the value proposition and had the win themes committed to memory. And it went well.

When you lose, it's hard. No one wants to lose. Usually the primary problem is lack of a close customer relationship. You weren't connecting with and talking to the right people or you didn't grasp what the right people were trying to tell you. Perhaps you got off to a late start and let the competitors shape the game. You might have failed to understand the customer decision process or hadn't the means to resolve internal teaming issues. And you may not have asked for an independent review of your solution and messages before you submitted your proposal.

If you don't share the best practices of lessons learned, you'll keep making the mistakes of lessons experienced. Here's hoping you'll wake up tomorrow and it's not still today.

Lesson Quotes

"The difference between school and life? In school, you're taught a lesson and then given a test. In life, you're given a test that teaches you a lesson."
--Tom Bodett

"They can conquer who believe they can. He has not learned the first lesson in life who does not every day surmount a fear."
--Ralph Waldo Emerson

"I have learned through bitter experience the one supreme lesson to conserve my anger, and as heat conserved is transmitted into energy, even so our anger controlled can be transmitted into a power that can move the world."
--Mahatma Gandhi

"When you give a lesson in meanness to a critter or a person, don't be surprised if they learn their lesson"
--Will Rogers

"Trust the people ... that is the crucial lesson of history."
--Ronald Reagan

"That's the reason they're called lessons, because they lesson from day to day."
--Lewis Carroll

"The charm of history and its enigmatic lesson consist in the fact that, from age to age, nothing changes and yet everything is completely different."
--Aldous Huxley

"A tough lesson in life that one has to learn is that not everybody wishes you well."
--Dan Rather

"The greatest lesson in life is to know that even fools are right sometimes."
--Horace

"He neither drank, smoked, nor rode a bicycle. Living frugally, saving his money, he died early, surrounded by greedy relatives. It was a great lesson

to me."
--John Barrymore

"I think the one lesson I have learned is that there is no substitute for paying attention."
--Diane Sawyer

"A good lesson in keeping your perspective is: Take your job seriously but don't take yourself seriously."
--Thomas P. "Tip" O'Neill

"I knew after my first lesson what I wanted to do with my life."
--Billie Jean King

"I shall the effect of this good lesson keep as watchman to my heart."
--William Shakespeare

"I've learned the lesson that when you're in the middle of something that seems overwhelming, or you're in a bad situation and it seems like it's the end of the world or whatever, then you learn that it's not."
--Lee Ann Womack

"If there is one lesson for U.S. foreign policy from the past 10 years, it is surely that military intervention can seem simple but is in fact a complex affair with the potential for unintended consequences."
--Fareed Zakaria

"If you must hold yourself up to your children as an object lesson, hold yourself up as a warning and not as an example."
--George Bernard Shaw

"Men of sense often learn from their enemies. It is from their foes, not their friends, that cities learn the lesson of building high walls and ships of war."
--Aristophanes

"One lesson you better learn if you want to be in politics is that you never go out on a golf course and beat the President."
--Lyndon B. Johnson

"The first lesson is that you can't lose a war if you have command of the air, and you can't win a war if you haven't."
--Jimmy Doolittle

"The important thing is to learn a lesson every time you lose. Life is a learning process and you have to try to learn what's best for you. Let me tell you, life is not fun when you're banging your head against a brick wall all the time."
--John McEnroe

"The lesson is that you can still make mistakes and be forgiven."
--Robert Downey, Jr.

"The most important lesson I've learned in this business is how to say no. I have said no to a lot of temptations, and I am glad I did."
--Penelope Cruz

"The show doesn't drive home a lesson, but it can open up people's minds enough for them to see how stupid every kind of prejudice can be."
--Redd Foxx

"This is the lesson that history teaches: repetition."
--Gertrude Stein

"What kind of people do they think we are? Is it possible they do not realize that we shall never cease to persevere against them until they have been taught a lesson which they and the world will never forget?"
--Winston Churchill

"You just have to keep on doing what you do. It's the lesson I get from my husband; he just says, Keep going. Start by starting."
--Meryl Streep

Listening

Do you listen to your customers? Really listen? I sometimes have trouble listening because I love to talk. I dive into conversations and steer them to fun places. A pause in someone's commentary is an opening for me to insert something witty. The world of words is so rich and I have so many of them I want to share. It has taken me years and a lot of self-discipline to slow down, stop talking and start listening.

Donald Weiss, a training and development executive and consultant, has some suggestions for poor listeners like me. In his book, *How to Deal with Difficult People*, he listed a number of common listening errors people make. The first set of two involves the mind: failure to concentrate or listening too hard. Almost all of us can hear something. Not all of us listen. Listening is an action that we can do too little or too much of. The second set of two common listening errors involves pacing: jumping the gun or lagging behind. Sometimes we miss nuance when we anticipate what the customer will say next and sometimes we fail to keep up with the pace of the conversation and make the logical connections the customer expects us to.

The third set of two common listening errors can be elusive to detect and correct: omitting or adding. I recall a business meeting I had in Brussels with a potential partner. My corporate contact, who knew the businessman, accompanied me on the visit, turning a cold call into a warm encounter. In my mind, the meeting went extraordinarily well. I was back at the hotel writing up my trip report to management and before I hit send I passed it over to my corporate colleague for his comment. He came back immediately: "Were we at the same meeting? He didn't say *yes*. He said *maybe if* ..." I was shocked to realize what I had done. I had cherry-picked all the positive comments, added a few flourishes, and overlooked all the guarded and qualified comments. I so much wanted the businessman to say yes to partnership that my mind constructed an alternative reality. Chastened, I rewrote the report and sent it back. We never cut the deal, but my colleague saved me from looking foolish. I learned several valuable lessons, including listening to teammates as well as customers.

Weiss also has a good list of customer-focused listening techniques, which I've adapted and adopted:

1. Give full attention – ignore distractions
2. Have an open mind – bad news is valuable
3. Ask questions – then ask some more questions
4. Acknowledge the customer's words and feelings
5. Make no excuses – avoid being defensive
6. Work out a mutually beneficial plan of action
7. Thank your customer for sharing insights

Regarding the first technique, I remember an office visit with a new deputy minister of defense for acquisition in a European country. My local consultant and I had finally gotten in to see him. We were escorted to sofas around a coffee table. The minister came in at last and sat behind his desk, putting distance and a barrier between us. Not a great start, so I thought to begin slow and try to listen. He was sitting by an open window and I realized he'd been cleaning his glasses, for there was a thin strip of toilet tissue hanging down from the left hinge of his frames. The room was warm. I was jet-lagged. My eyes followed the tissue as it moved back and forth in the breeze. To this day I have no idea what that man said. Fortunately, I was not alone and my consultant received the information we were seeking. After that I became much better at ignoring distractions and focusing on what the customer said and meant. But I still brought a business colleague to meetings whenever I could.

Listen Quotes

"Courage is what it takes to stand up and speak; courage is also what it takes to sit down and listen."
--Winston Churchill

"To listen is an effort, and just to hear is no merit. A duck hears also."
--Igor Stravinsky

"It is the province of knowledge to speak. And it is the privilege of wisdom to listen."
--Oliver Wendell Holmes

"Bore, n.: A person who talks when you wish him to listen."
--Ambrose Bierce

"It's a rare person who wants to hear what he doesn't want to hear."
--Dick Cavett

"Just because I didn't do what you told me, doesn't mean I wasn't listening to you!"
--Hank Ketcham

"Listening looks easy, but it's not simple. Every head is a world."
--Cuban Proverb

"Who speaks, sows; Who listens, reaps."
--Argentine Proverb

"If speaking is silver, then listening is gold."
--Turkish Proverb

"The greatest compliment that was ever paid me was when one asked me what I thought, and attended to my answer."
--Henry David Thoreau

"Listening, not imitation, may be the sincerest form of flattery."
--Dr. Joyce Brothers

"No man ever listened himself out of a job."
--Calvin Coolidge

"When people talk, listen completely. Most people never listen."
--Ernest Hemingway

"Your modern teenager is not about to listen to advice from an old person, defined as a person who remembers when there was no Velcro."
--Dave Barry

"I only wish I could find an institute that teaches people how to listen. Business people need to listen at least as much as they need to talk."
--Lee Iacocca

"The best salespeople are great listeners—that's how you find out what the buyer wants."
--Larry Wilson

"I remind myself every morning: Nothing I say this day will teach me anything. So if I'm going to learn, I must do it by listening."
--Larry King

"You learn when you listen. You earn when you listen—not just money, but respect."
--Harvey Mackay

"You have to master not only the art of listening to your head, you must also master listening to your heart and listening to your gut."
--Carly Fiorina

"Guitar players never listen to lead singers."
--Steven Tyler

"The key to success is to get out into the store and listen to what the associates have to say. It's terribly important for everyone to get involved. Our best ideas come from clerks and stockboys."
--Sam Walton

"Education is the ability to listen to almost anything without losing your temper or your self-confidence."
--Robert Frost

"A pessimist is a person who has had to listen to too many optimists."
--Don Marquis

"Be grateful for luck. Pay the thunder no mind - listen to the birds. And don't hate nobody."
--Eubie Blake

"Listen to many, speak to a few."
--William Shakespeare

"We have two ears and one mouth so that we can listen twice as much as we speak."
--Epictetus

"Never lose sight of the need to reach out and talk to other people who don't share your view. Listen to them and see if you can find a way to compromise."
--Colin Powell

"If you wish to know the mind of a man, listen to his words."
--Johann Wolfgang von Goethe

"Talk to a man about himself and he will listen for hours."
--Benjamin Disraeli

"Any problem, big or small, within a family, always seems to start with bad communication. Someone isn't listening."
--Emma Thompson

"The sun, the darkness, the winds are all listening to what we have to say."
--Geronimo

"I'm always relieved when someone is delivering a eulogy and I realize I'm listening to it."
--George Carlin

"If A equals success, then the formula is A equals X plus Y and Z, with X being work, Y play, and Z keeping your mouth shut."
--Albert Einstein

"Don't underestimate the value of Doing Nothing, of just going along, listening to all the things you can't hear, and not bothering."
--Winnie the Pooh

MAKE *A SMILE* - *SALES* MANAGEMENT

Make *a Smile*

Have you ever noticed how some people fill up the room with positive energy? More often than not this includes a warm smile. Considering the gravity of many of the projects you're involved in with government customers, this might seem a little shallow. But the goal is to gain trusted partnerships with your customers, right? And there's no better gateway to trust between human beings than the smile.

In a past issue of *Smithsonian Magazine* there was a fascinating article about the science of smiling. More than 170 years ago a Paris physician named Guillaume Duchenne was able to demonstrate with live electrodes placed on the human face the mechanistic nature of human facial expressions. The experiment proved that certain facial expressions – especially the smile – are innate, built into our biology long before language itself. These long ago findings are now widely accepted. Consider that athletes, blind from birth, smile exactly like "seeing" athletes at the moment of victory. It's why, on seeing a smile, our instinct is to smile back, before it's even possible for the conscious brain to be aware of the emotion being expressed. We really are hard-wired to smile.

But there's more to it than just reflex action. A recent study in Britain found that merely seeing a child's smile produces an emotional high. It's also known that it's practically impossible to think a bad thought and smile at the same time. Call center personnel for years have placed mirrors in front of themselves to monitor their faces because smiles generate more thoughtful responses and more sincerity to the tone of voice. A smile by itself triggers a positive feeling in the brain and people who smile a lot tend to be happier and more successful.

Of course, there are cultural variations. The Japanese, for example, are said to smile less often because of social tradition discouraging emotional displays. There also are smiles of embarrassment, love, contempt, pride, submission, flirtation and even polite tolerance.

It's the genuinely warm smile that's important to us. Remember Microsoft Office's smiling cartoon paper clip helper "Clippy," who drove us all crazy? Clippy didn't have the emotional intelligence to stop smiling when we were frustrated and we deleted him as soon as he told us how. And haven't we all been annoyed by the creepy smiles of service personnel who we know just don't care? A real smile is hard to fake. The cheek muscles lift the corners of the mouth, the muscles around the eyes contract, causing crow's feet wrinkling, often with the eyebrows drawing down a little. It sounds simple, but it's difficult to fake the eye muscle movement. All of us can sense when a smile isn't real.

The bottom line is this: the smile is humankind's most potent symbol of cooperation. Used appropriately and delivered genuinely, a smile can be an effective ally in gaining customer intimacy and trust.

Smile Quotes

"Look back, and smile on perils past."
--Walter Scott

"Don't cry because it's over. Smile because it happened."
--Dr. Seuss

"The real man smiles in trouble, gathers strength from distress, and grows brave by reflection."
--Thomas Paine

"Be thou the rainbow in the storms of life. The evening beam that smiles the clouds away, and tints tomorrow with prophetic ray."
--Lord Byron

"A smile is a curve that sets everything straight."
--Phyllis Diller

"Luck is not chance, it's toil; fortune's expensive smile is earned."
--Emily Dickinson

"If you smile when no one else is around, you really mean it."
--Andy Rooney

"Wear a smile and have friends; wear a scowl and have wrinkles."
--George Eliot

"Children learn to smile from their parents."
--Shinichi Suzuki

"A smile abroad is often a scowl at home."
--Alfred Lord Tennyson

"A kind heart is a fountain of gladness, making everything in its vicinity freshen into smiles."
--Washington Irving

"A smile is the chosen vehicle of all ambiguities."
--Herman Melville

"I think that anybody that smiles automatically looks better."
--Diane Lane

"She gave me a smile I could feel in my hip pocket."
--Raymond Chandler

"If you have only one smile in you, give it to the people you love."
--Maya Angelou

"We shall never know all the good that a simple smile can do."
--Mother Teresa

"It takes a man to suffer ignorance and smile."
--Sting

"If you're able to help some people and make them smile and make them realize that life is good, then that's worth so much more than buying a pair of shoes."
--Maria Sharapova

"Smile in the mirror. Do that every morning and you'll start to see a big difference in your life."
--Yoko Ono

"Start every day off with a smile and get it over with."
--W. C. Fields

"War is a game that is played with a smile. If you can't smile, grin. If you can't grin, keep out of the way till you can."
--Winston Churchill

"Everyone looks so much better when they smile."
--Jimmy Fallon

"Let me smile with the wise, and feed with the rich."
--Samuel Johnson

"The man who can smile at his breaks and grab his chances gets on."
--Samuel Goldwyn

"This is a time when we need to smile more and Hollywood movies are supposed to do that for people in difficult times."
--Steven Spielberg

"A stale article, if you dip it in a good, warm, sunny smile, will go off better than a fresh one that you've scowled upon."
--Nathaniel Hawthorne

"You know when someone's over-flattering you in a way. You smile but you can't believe it."
--Laura Linney

"I think Led Zeppelin must have worn some of the most peculiar clothing that men had ever been seen to wear without cracking a smile."
--Robert Plant

"If you smile things will work out."
--Serena Williams

"May Heaven be propitious, and smile on the cause of my country."
--Zebulon Pike

"Nothing you wear is more important than your smile."
--Connie Stevens

"When you see that many people with a smile on their face, then you must be doing something right."
--Greg Norman

"You get somebody to crack a smile, that's a beautiful thing."
--Tracy Morgan

"One who smiles rather than rages is always the stronger."
--Japanese Proverb

"The robb'd that smiles, steals something from the thief."
--Shakespeare

"You've got to get up every morning with a smile on your face ..."
--Carole King

Sales Management

Why is it that so few in the aerospace, defense and technology sectors discuss sales management? Is it because people in business development don't think they're selling? Or since business development isn't a technical area it doesn't need to be managed? I don't really know. When I first came into business development I was in awe of the professionals who put together capture campaigns. I studied and learned the detailed capture processes to help pull vast resources and large numbers of people together to gain contracts for eye-watering technical solutions. But along the way I was struck by how we could complete all the checklists and still lose to competitors. We didn't appear to have done anything wrong - we just didn't win. The reasons given for this were usually "the customer made a mistake in not choosing us" or "the competition didn't play fair." Could at least part of the problem have been that we didn't close a deal because we weren't managing our sales force well? If you think that might be a possibility in your company, here are some actions to consider for managers of capture leads and their teams.

Select

Not everyone is cut out to be in sales. John Asher of Asher Strategies says that about 20 percent of us are naturally suited to it and could become super salespeople. About 60 percent of us are trainable and capable of being effective. And the remaining 20 percent should never be put in front of a customer. You're probably smiling now because you know someone who fits that last category. Those people can be extremely talented and have great value to the company – just not in customer-facing roles. Most people arrive into government-oriented business development internally, up from the programs, or externally, over from the customer environment. They might have great knowledge and experience, but are their personalities suited to sales? Are they proactive and people-oriented with a positive disposition and endless patience? Or are they more comfortable going to the office every day and plotting strategies and organizing support programs (skills better suited to sales managers than sales staff)? When you hire someone to sell your programs, do you consider "what" they are in addition to "who" they are?

Train

If you're in business development, have you ever been given the opportunity to take a sales course? All great and successful companies train their sales people. Sales techniques are nothing new, but they're not common knowledge, either. You've taken lessons to improve your golf game, right? If you're a sales manager could it be a good idea to train your people how to outsell the competition? Or do you think your products are so good they sell themselves? There are a number of basic sales courses available that teach fundamentals applicable to any sales process, whether you are selling front line fighters or back office IT support. My colleagues listed at the back of this book can help you find the right course to help your business developers get the right skills. If you want to win, you need to train your team.

Equip

What do your business developers have in the way of equipment? Cell phones and laptops are essentials, but you might want to consider some other sales enablers. Could customized smart phone apps make your sales staff smarter? What about tablets? One company I've worked with presented each of its program managers a tablet loaded with other program offerings for on-contract-growth, so they could cross-sell and up-sell to current customers. This was a great idea, though it would have been even greater if they'd taught their program managers the fundamentals of sales. Tablets are changing the sales dynamic because they allow personalized, desk-side video demonstrations of products and services, showing what your solutions can "do," not just what they "are." And what about customer relationship management (CRM) tools to track your customers so you can qualify leads, deconflict meetings and keep your promises? Better tools = better profits.

Manage

Think about the best managers you ever had. Workplace stress is caused more by managers than tasks. Likely your best managers were the ones who made it easier for you to do your job. Administrative tasks pile up in organizations over time. What's the administrative workload of your best sales people? Are they spending more time working on reports than working on the customers? Good sales managers enable – not disable – their sales staff. Perhaps you could add more internal sales staff, streamline processes, or even do away with processes that no longer have purpose. Give your sales people the gift of time and they will reward you by meeting their stretch goals.

Mentor

If you have years of experience in business development, be generous and invest in your future sales force. Most up-and-coming business developers would rather talk about career development than sales development, but there is no better way to help them advance than by teaching them how to pull in big contracts. Unfortunately, mentoring is hard for many super sales people because they tend to work alone (and instinctively) and have little patience with people who don't "get it" right away. If you're a sales manager, don't just advise your people; inspire them. Make sure the newcomers to your organization team up with the right veterans and have the opportunity to go on sales calls with them. Everyone will profit, and so will your line of business.

Management Quotes

"Every manager knows how to count, but smart managers know what counts."
--Anonymous

"Next to knowing all about your own business, the best thing is to know all about the other fellow's business."
--John D. Rockefeller

"The successful man is the one who finds out what is the matter with his business before his competitors do."
--Roy L. Smith

"Your competition is EVERYTHING else your prospect could conceivably spend their money on."
--Don Cooper

"A successful man is one who can lay a firm foundation with the bricks others have thrown at him."
--David Brinkley

"Right now, this is a job. If I advance any higher, this would be my career. And if this were my career, I'd have to throw myself in front of a train."
--Jim Halpert

"Success means only doing what you do well, letting someone else do the rest."
--Goldstein S. Truism

"I'm not the smartest fellow in the world, but I can sure pick smart colleagues."
--Franklin D. Roosevelt

"Make sure you have a vice president in charge of your revolution, to engender ferment among your more conventional colleagues."
--David Ogilvy

"Good management is the art of making problems so interesting and their solutions so constructive that everyone wants to get to work and deal with them."
--Paul Hawken

"Management is efficiency in climbing the ladder of success; leadership determines whether the ladder is leaning against the right wall."
--Stephen R. Covey

"A good manager is a man who isn't worried about his own career but rather the careers of those who work for him."
--H. S. M. Burns

"Get the right people on the bus and in the right seat."
--Jim Collins

"Management is doing things right; leadership is doing the right things."
--Peter F. Drucker

"Most people do not receive nearly enough appreciation. How can this be when appreciation is free, easy, and readily available? All you have to do is speak. Go give some away now."
--Rhoberta Shaler

"Early to bed and early to rise probably indicates unskilled labor."
--John Ciardi

"By working faithfully eight hours a day you may eventually get to be boss and work twelve hours a day."
--Robert Frost

"Goals allow you to control the direction of change in your favour."
--Brian Tracy

"Goals aren't enough. You need goals plus deadlines: goals big enough to get excited about and deadline to make you run. One isn't much good without the other, but together they can be tremendous."
--Ben Feldman

"Men and women want to do a good job, a creative job, and if they are provided the proper environment, they will do so."
--Bill Hewlett

"There is no limit to what a man can do so long as he does not care a straw who gets the credit for it."
--Charles Edward Montague

"Trust—the glue that binds followers and leaders together."
--Warren Bennis

"The trust that we put in ourselves makes us feel trust in others."
--François de la Rochefoucauld

"People want to feel what they do makes a difference."
--Frances Hesselbein

"If you want someone to do a good job, give them a good job to do."
--Mike Abrashoff

"Obstacles can't stop you. Problems can't stop you. Most of all, other people can't stop you. Only you can stop you."
--Jeffrey Gitomer

"Great works are performed not by strength but by perseverance."
--Samuel Johnson

"Take risks. If you win, you'll be happy; if you lose, you'll be wise."
--Anonymous

"Not failure, but low aim, is crime."
--James Russell Lowell

"If at first you don't succeed; you are running about average."
--MH Alderson

"If at first you don't succeed, take the tax loss."
--Kirk Kirkpatrick

SAYING NO

Saying No

It's difficult to say "No" to a customer. There's a classic story involving former Southwest Airlines CEO Herb Kelleher, who instilled a corporate culture of employees taking their jobs seriously, but themselves not so seriously. A cranky customer wrote directly to Herb giving great detail on what she didn't like about his airline's service. Herb wrote back, "Dear Mrs. Smith, We will miss you." In the aerospace, defense and technology business you often run into cranky customers and sometimes their demands seem unreasonable and unbearable. Unfortunately, there are few of them you could write a "Herb" to. They are interlocked with the rest of the customer set and you need them to keep your business alive.

A woman who worked for me several years ago spoke at length about her adventures running a small shop in Brussels. Good help was hard to find and she struggled to encourage her store employees to engage with customers in more positive ways. Eventually she placed a sign near the cash register where it could only be read by employees ringing up sales. The sign said: "THE CUSTOMER IS ALWAYS RIGHT *sometimes confused, mis-informed, rude, stubborn, changeable, and even downright stupid* BUT NEVER WRONG." That's the other end of the scale from Herb's letter. Such an approach may work okay in a commercial store setting, but what about selling to government customers, where they control much of your business and can be most of the problem?

Recall the section on Handling Difficult Customers. The professionals who do call center work advise that before you can fix a problem situation you must first understand the customer and then control the emotion. Try to say "No" positively. First state why you can't say yes (legal concerns, policies/procedures, immoral/unethical, not physically possible) and then provide pertinent information and follow up quickly by telling them what you CAN do. Saying no is not an end in itself, but can be the beginning of a dialogue leading to a better relationship with the customer.

The above is especially good advice if you have a communication or technical breakdown with an existing customer. But what about those customers more at the front of the business cycle, where you are shaping new business or shortly after you've gained new business? What about government customers whose unrestrained requests are leading to "requirements creep" with the potential to doom the programs you've fought so hard to shape and win? How do you say no to them?

Professor Jeffrey Williams of Carnegie Mellon University's Tepper School of Business knows the aerospace, defense, and technology business sector well, having worked as an engineer in the Apollo program before turning to academia. Once, in a presentation, he challenged a group of business leaders to "learn to say no and make 'em like it." If you don't, he argued, you can't effectively manage the business case. If you don't tell your customers "no" to certain demands, you cannot establish boundaries for performance and put standards in place to remain competitive in your markets. I later asked Professor Williams if he could expand on his challenge. He acknowledged that this is a tough area and "a chapter I've yet to write." It's still a great teaching point, however, especially the "Make 'Em Like It" part.

Soon after that conversation I came across a book I'd forgotten about: William Ury's *The Power of a Positive No*. Ury is cofounder of Harvard's Global Negotiation Project and author of the iconic *Getting to Yes*. In the opening pages of *The Power of a Positive No*, Ury writes: "Perhaps the single biggest mistake we make when we say No is to *start* from No. We derive our No from what we are *against* – the other's demand or behavior. In contrast, a Positive No calls on us to base our No on what we are *for*. Instead of starting from No, start from Yes – a Yes to your core interests and what truly matters."

There's a tendency to think this is too hard, especially in the aerospace, defense and technology sector, where there are only a few customers and the stakes are so high. Because you are reluctant to say No, you tend to take one of three actions: accommodate by saying Yes when you really want to say No, attack by saying No poorly and triggering a conflict, or avoid by not addressing the issue at all. Ury refers to this as the *Three-A Trap: Accommodate, Attack, Avoid.*

When faced with saying a difficult No, Ury recommends the formula *Yes No Yes*. The first *Yes* is a reminder to reflect on what is in your best interest. This grounds your subsequent *No* in something positive that is important to you. Take some time to make sure your *No* is stated in a respectful way, yet still centered on the self-interests you are trying to protect. After you've said your committed and respectful *No*, it's time to suggest an alternative and perhaps divert the customer's impossible suggestion. This is the second *Yes* - a positive request to consider another solution. Diplomats have been doing this for centuries.

In a review of Ury's book, *Time* magazine's Barbara Kiviat illustrated the concept with a story. "John Kenneth Gailbraith's housekeeper never had a problem saying no. One day President Lyndon Johnson called the Galbraith house wanting to talk to the great economist. 'He's taking a nap and has left strict orders not to be disturbed,' said the housekeeper. Johnson replied, 'Well, I'm the President. Wake him up.' The response: 'I'm sorry, Mr. President, but I work for Mr. Galbraith, not for you.' Click." Kiviat went on to explain that Ury's *Yes No Yes* formula, properly applied, can give a huge two-fold payoff. Providing a respectful, but decisive no can fix the problem and actually strengthen the customer relationship. "Just consider what happened when Galbraith woke from his nap and returned Johnson's call. 'Who is that woman?' the President asked, inquiring about the housekeeper who had dared tell him no. 'I want her working for me'."

Saying No Quotes

"Customers trust you more if you say 'no' when the answer is no."
--Alexander Kjerulf

"Half of the troubles of this life can be traced to saying yes too quickly and not saying no soon enough."
--Josh Billings

"The art of leadership is saying no, not saying yes. It is very easy to say yes."
--Tony Blair

"The film industry is about saying 'no' to people, and inherently you cannot take 'no' for an answer."
--James Cameron

"[Success] comes from saying no to 1,000 things to make sure we don't get on the wrong track or try to do too much."
--Steve Jobs

"You have to decide what your highest priorities are and have the courage - pleasantly, smilingly, non-apologetically - to say 'no' to other things. And the way you do that is by having a bigger 'yes' burning inside."
--Stephen Covey

"Just Say No."
--Nancy Reagan

"Practice makes perfect. Saying 'no' as often as you can is a great way to get better at it and more comfortable with saying the word."
--Leo Babauta

"I hate a man who always says 'yes' to me. When I say 'no' I like a man who also says 'no'."
--Samuel Goldwyn

"What is a rebel? A man who says no."
--Albert Camus

"Yes and no are very short words, but we should think for some time before saying them."
--Anonymous

"No ham, no fowl."
--Apu Nahasapeemapetilon *(The Simpsons)*

"The other day, my little boy talked back to my wife. She told him to do something; he said, 'No, I don't want to.' So, I had to pull him aside and say, 'Listen -- you gotta teach me how to do that'."
--Brian Kiley

"Yes! We have no bananas."
-- Eddie Cantor

"A 'No' uttered from the deepest conviction is better than a 'Yes' merely uttered to please, or worse, to avoid trouble."
--Mahatma Gandhi

"...there are often many things we feel we should do that, in fact, we don't really have to do. Getting to the point where we can tell the difference is a major milestone in the simplification process."
--Elaine St. James

"No is a complete sentence and so often we forget that. When we don't want to do something we can simply smile and say no. We don't have to explain ourselves. I found developing the ability to say no expanded my ability to say yes and really mean it."
--Susan Gregg

"Tone is the hardest part of saying no."
--Jonathan Price

"Being unable to say no can make you exhausted, stressed and irritable."
--Auliq Ice

"Information overload (on all levels) is exactly WHY you need an "ignore list". It has never been more important to be able to say, 'No'."
--Mani S. Sivasubramanian

"Say no to everything, so you can say yes to the one thing."
--Richie Norton

"Learn to say 'no' to the good so you can say 'yes' to the best."
--John C. Maxwell

"I only have 'yes' men around me. Who needs 'no' men?"
--Mae West

"I went down the street to the 24-hour grocery. When I got there, the guy was locking the front door. I said, 'Hey, the sign says you're open 24 hours.' He said, 'Yes, but not in a row'."
--Steven Wright

"Never allow a person to tell you no who doesn't have the power to say yes."
--Eleanor Roosevelt

OBJECTIONS TO PRICE - OPPORTUNITIES

Objections to Price

There are five words we all hate to hear from a customer: "Your price is too high." The first thing to remember when you hear this is don't be even the slightest bit defensive. That's a customer turnoff that can make you appear arrogant. A second thing to consider is that you may not have differentiated yourself enough from the competition. Have you leveled the playing field against competing products? And shown your customer three main features of your product or solution that (1) have value to the customer, (2) are easy to defend, and (3) are unique to you? (Power Messaging by Corporate Visions) If you haven't, then you may be in a cost shoot-out competition, something to avoid if at all possible.

Your government customers are couching price objections in terms of "controlling costs." Their budgets aren't what they used to be, but their needs are still great. Of course, the enlightened government customer understands that cost control is a two-way street with contractors, not just another opportunity to squeeze down margins. But regardless of what kind of government customers you have, they are capable of understanding value. And value is the answer to an objection on price.

In business surveys over the past decade, price was rarely in the top five lists of customer priorities. The major customer concerns have been a company's reputation (for delivery on time, in budget and in scope), the total solution provided, the quality of a company's personnel and their grasp of the customer's business needs. It's true that the global economic environment has changed for the worse, but you don't sell stuff that thrifty homeowners economize on. You're all about government security. While there may be more restrained requirements in future RFPs, there really are no restraints on threats. Money will be spent on security. Will it be spent with you? With dwindling budgets and growing dangers, you need to help your customers creatively and economically meet their security challenges. And to do that you need value-oriented offerings.

If you get a price objection from a government customer, it could be that they really want your product, but your price simply exceeds their budget. If that's the case you can propose alternative payment structures, look for shared funding resources, rescope your offering, reduce fidelity/resolution or another area of robustness, prioritize needs and fund the most important first (see *Non-manipulative Selling* by Tony Alessandra for further ideas).

If you know your price is going to be higher than the competition, bring it up before the customer does. In most cases, the buyer will appreciate your straightforwardness. This can then set you up to take the initiative in explaining the greater value behind your higher price. In this way you can build trust along with understanding and shorten the buying cycle.

The best way to help your customer see the full value of your offering is by giving them a good return on investment analysis. This is especially true for executive level "economic" buyers. The good ones are obsessed with ROI. The best ROI delivery techniques I've seen are in *How to Become a Rainmaker* by Jeffrey J. Fox:

- Express ROI in the clearest financial terms. Most purchases are made to solve problems ("pain") or grow capacity ("gain")
- Justify your offering's price by demonstrating how quickly the offering will pay for itself
- Show the buyer how taking no action or selecting an inferior solution costs more than funding your proposal
- Use the information as themes in proposals, listing both top-line improvers and bottom-line improvers
- Get the ROI in front of key decision makers as soon as possible

Also try using the Benson and Karasik "feel, felt, found" formula to communicate ROI in response to "Your price is too high":

- "I understand how you *feel*."
- "Some of our other customers *felt* the same way when they first heard of our solution."

- "But once they implemented our solution, they <u>found</u> that they got a substantial ROI, such as..."

This is a great technique that moves the customer from an emotional state to a factual state (for more great advice see *22 Keys to Sales Success* by James Benson and Paul Karasik).

The classic sales aphorism that reminds us to emphasize benefits over features is "Sell the sizzle, not the steak." The sizzle is the ROI and that's the best way to show value over price.

Price Quotes

"Price is what you pay. Value is what you get."
--Warren Buffett

"What is a cynic? A man who knows the price of everything and the value of nothing."
--Oscar Wilde

"Hard work is the price we must pay for success. I think you can accomplish anything if you're willing to pay the price."
--Vince Lombardi

"No price is too high to pay for the privilege of owning yourself."
--Friedrich Nietzsche

"The price of anything is the amount of life you exchange for it."
--Henry David Thoreau

"You do not pay the price of success, you enjoy the price of success."
--Zig Ziglar

"Where quality is the thing sought after, the thing of supreme quality is cheap, whatever the price one has to pay for it."
--William James

"There is hardly anything in the world that some man cannot make a little worse and sell a little cheaper, and the people who consider price only are this man's lawful prey."
--John Ruskin

"Fortune is like the market, where, many times, if you can stay a little, the price will fall."
--Francis Bacon

"A good film is when the price of the dinner, the theatre admission and the babysitter were worth it."
--Alfred Hitchcock

"People want economy and they will pay any price to get it."
--Lee Iacocca

"All wish to possess knowledge, but few, comparatively speaking, are willing to pay the price."
--Juvenal

"I know the price of success: dedication, hard work, and an unremitting devotion to the things you want to see happen."
--Frank Lloyd Wright

"The bitterness of poor quality is remembered long after the sweetness of low price has faded from memory."
--Aldo Gucci

Opportunities

Have you ever wondered about missed opportunities? Even the most successful people have looked back and wished they had recognized a great opportunity and acted in time. It's probably the hardest thing we face in life and business: to separate true opportunities from risky choices and dead ends.

People who lead successful professional and personal lives develop strategies and stick to them. But sometimes you can be so fixated on your strategies and daily processes that you overlook golden opportunities sitting right in front of you. While some of these opportunities could speed achievement of your strategic goals, they may not take you there in a direct line, which in turn cause you to hesitate. Think back on your career. Was there ever a time when you didn't get a job that you thought was perfect? And that disappointing outcome led you to an even better job? The same thing can happen with a small, seemingly inconsequential business opportunity - easy to overlook, yet could lead to a franchise for the enterprise if pursued with purpose.

Opportunities are best seen if you keep your eyes wide open. The concept of "wide eyes" is as old as our ancestors hunting in the forests and paying attention to the minutest of movements with their peripheral vision. The hunters that perfected this technique had more opportunities to take down game and therefore ate better and thrived. Here's a more modern example of "wide eyes." When you were first learning to drive, you quickly learned that if you focused too closely on the road immediately in front of the hood you tended to go slower and overcorrect the steering. But if you looked up toward the horizon and drove with "wide eyes," you were smoother, more assured in your steering and got to your destination faster (and with fewer surprises in your path).

Let's say in business discussions your customer reveals an unmet need that's causing big problems. Your customer may be dropping you a hint or indirectly asking for help. What do you do? Perhaps the customer need doesn't seem to be immediately compatible with what you're selling or the contract you're performing on. It's likely you're busy and want to move on, so you nod your head and switch the subject. But pause and ask yourself the classic questions: Is it a real opportunity? Is it on strategy? Can we win it? And is it worth it? If you approach the answers to those questions honestly and with "wide eyes," you might be surprised that your customer just gave you an opportunity to grow your business. But is it a good opportunity?

Business Development leaders can help their teams discern good opportunities from bad with proven sales processes. According to a recent Sales Benchmark Index, companies that deploy formal sales processes win 48 percent more business and shorten their sales cycles by 37 percent. Your company may have terrific business capture processes that, when followed, greatly improve your success in winning good new business from qualified opportunities. But when faced with many choices of where to invest your time and resources, it's often hard to determine which opportunity will lead to good new business.

Many successful companies use opportunity assessment tools as sales aids. These tools, and there are many of them readily available in template form, force you to ask hard questions about the state of an opportunity. They can be tailored to any business sector and are opportunity-specific internal aids intended to be maintained by sales managers. Actions are assessed at each stage of the sales cycle (including competitor positions) and go/no-go decisions appear throughout. Companies that effectively use tools such as this start assessing opportunities early in their sales process (Target, Qualify, Investigate, Propose, Close) and keep the data current for all to see. Get an opportunity assessment tool, pick an opportunity your unit is working on and see if you can answer the questions. How many yes's and no's do you have? Does this cause you to reassess the opportunity or your approach to it?

This is not to say that you should rush to change your sales processes if they're working well for your line of business and customer set. But if you don't have an effective opportunity process (or any process at all), perhaps you should consider adopting one to systematically open your eyes to real opportunities and turn them into growth programs. Good hunting.

Opportunity Quotes

"Sometimes we stare so long at a door that is closing that we see too late the one that is open."
--Alexander Graham Bell

"Small opportunities are often the beginning of great enterprises."
--Demosthenes

"A pessimist sees the difficulty in every opportunity; an optimist sees the opportunity in every difficulty."
--Winston Churchill

"I was seldom able to see an opportunity until it had ceased to be one."
--Mark Twain

"When written in Chinese, the word 'crisis' is composed of two characters. One represents danger and the other represents opportunity."
--John F. Kennedy

"No great man ever complains of want of opportunity."
--Ralph Waldo Emerson

"We are confronted with insurmountable opportunities."
--Pogo (Walt Kelly)

"Failure is simply the opportunity to begin again, this time more intelligently."
--Henry Ford

"Luck is what happens when preparation meets opportunity."
--Seneca

"It pays to know the enemy – not least because at some time you may have the opportunity to turn him into a friend."
--Margaret Thatcher

"Statistics suggest that when customers complain, business owners and managers ought to get excited about it. The complaining customer represents a huge opportunity for more business."
--Zig Ziglar

"Ability is nothing without opportunity."
--Napoleon Bonaparte

"The reason a lot of people do not recognize opportunity is because it usually goes around wearing overalls looking like hard work."
--Thomas A. Edison

"The entrepreneur always searches for change, responds to it, and exploits it as an opportunity."
--Peter Drucker

"If opportunity doesn't knock, build a door."
--Milton Berle

"I always tried to turn every disaster into an opportunity."
--John D. Rockefeller

"You better lose yourself in the music, the moment. You own it. You better never let it go. You only get one shot. Do not miss your chance to blow. This opportunity comes once in a lifetime."
--Eminem

"To improve the golden moment of opportunity, and catch the good that is within our reach, is the great art of life."
--Samuel Johnson

"A man who misses his opportunity, and monkey who misses his branch, cannot be saved."
--Hindu Proverb

"Opportunity is lost by deliberation."
--Publilius Syrus

"He who refuses to embrace a unique opportunity loses the prize as surely as if he had failed."
--William James

"The man who grasps an opportunity as it is paraded before him, nine times out of ten makes a success, but the man who makes his own opportunities is, barring an accident, a sure-fire success."
--Dale Carnegie

"A wise man will make more opportunities than he finds."
--Francis Bacon

"Summing up, it is clear the future holds great opportunities. It also holds pitfalls. The trick will be to avoid the pitfalls, seize the opportunities, and get back home by six o'clock."
--Woody Allen

"Four things come not back. The spoken word, the sped arrow, the past life, and the neglected opportunity."
--Arabic Proverb

"The early bird may get the worm, but the second mouse gets the cheese."
--Stephen Wright

"Too often the opportunity knocks, but by the time you push back the chain, pull back the bolt, unhook the locks and shut off the burglar alarm, it's too late."
--Rita Coolidge

"Next to knowing when to seize an opportunity, the most important thing in life is to know when to forego an advantage."
--Benjamin Disraeli

"To hell with circumstances; I create opportunities."
--Bruce Lee

"We are continually faced by great opportunities brilliantly disguised as insoluble problems."
--Lee Iacocca

"Everyone has bad breaks, but everyone also has opportunities. The man who can smile at his breaks and grab his chances gets on."
--Samuel Goldwyn

"Let us my friends snatch our opportunity from the passing day."
--Horace

"Listen to the mustn'ts child. Listen to the don'ts. Listen to the shouldn't haves, the impossibles, the won'ts. Listen to the never haves, then listen close to me. Anything can happen child. Anything can be."
--Shel Silverstein

P

PAIN - PEOPLE - PERSONALITIES - *SALES* PROCESS - PROMISES

Pain

In shaping, capturing, and extending our business, it's vital to keep in mind what are called "pain points" in sales-speak. These could be unpleasant budget, performance, or schedule characteristics of a current product or service or they could be urgent unmet needs. More often than not, pain points aren't readily disclosed by the customer. They could involve political pressures and dysfunctional internal processes or they could be systemic aches that the customer has trouble defining. In all cases, the customer simply wants the pain to go away.

I recall a story told by Mike Miller of Corporate Visions. He was a runner and liked to listen to music while he ran. Early on he had a CD player that attached to a belt around his waist, but the CD skipped now and then. It was annoying. He thought the new "anti-skip" technology being developed would be a good fix. But then a company offered a whole new value proposition: "What if you could have a thousand songs in your pocket?" And he bought an iPod and couldn't believe how pleasant it became to run with music. Not only did the iPod not skip, it was light and he could change music anytime he wanted. No pain. And Apple delivered a franchise product that dominated the market. The manufacturers of now-outdated CD players delivered what the runner asked for – a CD that didn't skip – but that's not what he truly wanted.

How does the example above apply to the aerospace and defense industry? If you care about managing customer expectations and building franchise programs to grow your business it does. Your customers expect your products to be as easy to work with as their personal appliances. Think about it. Your government customers, especially in defense, have a lot more pain points than a recreational runner. They are dealing with the aftermath of multiple armed conflicts, military equipment inventories that are aging and needing replacement, a sputtering economy that is putting downward pressure on acquisition and RDT&E budgets, political leaders who have highly divergent opinions on defense postures, and many of their own employment positions are being eliminated and their compensation is being "adjusted." Ouch.

Consider the customer you want to reach out to and imagine what a day in his or her life might be like, especially in your area of business interest (see Day in the Life of a Customer in section D). Try to feel the pain points. In your mind, shift to the customer's perspective and look back at your company. What do you see? A company that just wants to sell you something? Or people who want to help you do your job better, faster and cheaper? A government contractor like all the others? Or friendly partners who can make your life easier by taking away the pain?

The thing is - it's not the pill you are selling that's really so important. It's the headache the customer wants to get rid of. And if you focus on helping customers get rid of their pain, you may start feeling better yourself.

Pain Quotes

"Pain is inevitable. Suffering is optional."
--Anonymous

"One can find so many pains when the rain is falling."
--John Steinbeck

"When there is pain, there are no words. All pain is the same."
--Toni Morrison

"The worst pain a man can suffer: to have insight into much and power over nothing."
--Herodotus

"Given the choice between the experience of pain and nothing, I would choose pain."
--William Faulkner

"Pain is such an uncomfortable feeling that even a tiny amount of it is enough to ruin every enjoyment."
--Will Rogers

"Painful as it may be, a significant emotional event can be the catalyst for choosing a direction that serves us-and those around us - more effectively. Look for the learning."
--Louisa May Alcott

"If pain could have cured us we should long ago have been saved."
--George Santayana

"Criticism may not be agreeable, but it is necessary. It fulfills the same function as pain in the human body. It calls attention to an unhealthy state of things."
--Winston Churchill

"History, despite its wrenching pain, cannot be unlived, but if faced with courage, need not be lived again."
--Maya Angelou

"How much pain they have cost us, the evils which have never happened."
--Thomas Jefferson

"You can judge your age by the amount of pain you feel when you come in contact with a new idea."
--Pearl S. Buck

"I assess the power of a will by how much resistance, pain, torture it endures and knows how to turn to its advantage."
--Friedrich Nietzsche

"One good thing about music, when it hits you, you feel no pain."
--Bob Marley

"The aim of the wise is not to secure pleasure, but to avoid pain."
--Aristotle

"There is a thin line that separates laughter and pain, comedy and tragedy, humor and hurt."
--Erma Bombeck

"To truly laugh, you must be able to take your pain, and play with it!"
--Charlie Chaplin

"It is easier to find men who will volunteer to die, than to find those who are willing to endure pain with patience."
--Julius Caesar

"You gotta love livin', baby, 'cause dyin' is a pain in the ass."
--Frank Sinatra

"The greatest evil is physical pain."
--Saint Augustine

"I'm not into working out. My philosophy: No pain, no pain."
--Carol Leifer

"There is no coming to consciousness without pain."
--Carl Jung

"Give me life, give me pain, give me myself again."
--Tori Amos

"The great art of life is sensation, to feel that we exist, even in pain."
--Lord Byron

"Pain is as diverse as man. One suffers as one can."
--Victor Hugo

"Men who have a pierced ear are better prepared for marriage - they've experienced pain and bought jewelry."
--Rita Rudner

"The two enemies of human happiness are pain and boredom."
--Arthur Schopenhauer

"My focus is to forget the pain of life. Forget the pain, mock the pain, reduce it. And laugh."
--Jim Carrey

"If it doesn't work out there will never be any doubt that the pleasure was worth all the pain."
--Jimmy Buffett

"A wise man will make haste to forgive, because he knows the true value of time, and will not suffer it to pass away in unnecessary pain."
--Samuel Johnson

"To hurt is as human as to breathe."
--J. K. Rowling

"Leadership is, among other things, the ability to inflict pain and get away with it – short-term pain for long-term gain."
--George Will

"Pain is temporary, film is forever."
--Michael J. Fox

People

It's been said that people want to buy from people, not from companies. It reminds me once again of the importance that people play in any business. Not just customers, but you and those connected with you. So as you contemplate the coming challenge of doing business tomorrow, next week, next year, perhaps it might be worth a look at two people-oriented business enablers: inside mentors and outside networks.

Mentors can make all the difference in your career and your life. If you've been lucky enough to have had a good mentor you'll know what I'm talking about. I still remember the day early in my career when I told my mentor I was doing everything I was supposed to be doing and yet my promotion status still looked uncertain. He replied: "You don't get promoted by solving problems related to your job description. That's what we hired you to do. You get promoted by solving your boss's problems." With three short sentences he set me on a path to success.

Your company may have formal initiatives to link employees to mentors. If that works for you, great. But sometimes it's hard to get the chemistry right. You might want to consider reaching out to a colleague to help find a mentor who's the right fit and willing to invest time in your success. In sales management studies it's been shown that business developers can increase their productivity up to 20 percent by being actively mentored and coached by experienced professionals. So in being open to mentoring (asking for help) and in being willing to serve as an active mentor (offering help), people can help each other and grow the business.

Outside professional networks also can help you grow professionally and increase your productivity. A normal reaction to adding something to your busy schedule is, "I don't have time. We're short staffed. There's no way." But there's never really enough time, is there?

You work on projects until you run out of time, often devoting precious minutes to perfecting solutions; sometimes violating business guru Tom Connellan's dictum that "There is no point in doing well that which you should not be doing at all."

The interesting thing about adding an outside professional network to your personal and business life is that it might even save you time, especially if you're in business development and the professional networks are linked to your customer base. For one thing, participating in a professional association (such as AIAA, AIA, AFA, etc.) will help you better understand the different aspects of your customer's environment – technology, economics and politics. Why pay a consultant when you can find out first hand? A second reason to expand your external network of professional associations is that you can pick up tips from discussions with industry peers. The more you talk with others in your field, the more you understand different ways to approach the same business opportunity. This widens the aperture of viewing your market and the competitive landscape. A third reason to join a professional association is that others will see you in a new light. In particular, your customers may begin to see you more as a peer than a salesperson. And that's exactly what you want.

(Some) People Quotes

"Some people regard private enterprise as a predatory tiger to be shot. Others look on it as a cow they can milk. Not enough people see it as a healthy horse, pulling a sturdy wagon."
--Winston Churchill

"Rarely do we find men who willingly engage in hard, solid thinking. There is an almost universal quest for easy answers and half-baked solutions. Nothing pains some people more than having to think."
--Martin Luther King, Jr.

"Some people die at 25 and aren't buried until 75."
--Benjamin Franklin

"Some people wonder all their lives if they've made a difference. The Marines don't have that problem."
--Ronald Reagan

"Be a yardstick of quality. Some people aren't used to an environment where excellence is expected."
--Steve Jobs

"Always do right. This will gratify some people and astonish the rest."
--Mark Twain

"Some people try to find things in this game that don't exist but football is only two things – blocking and tackling."
--Vince Lombardi

"There are some people who, if they don't already know, you can't tell 'em."
--Yogi Berra

"Some people see things that are and ask, 'Why?' Some people dream of things that never were and ask, 'Why not?' Some people have to go to work and don't have time for all that."
--George Carlin

"Some people want it to happen, some wish it would happen, others make it happen."
--Michael Jordan

"Some people think luxury is the opposite of poverty. It is not. It is the opposite of vulgarity."
--Coco Chanel

"Nobody realizes that some people expend tremendous energy merely to be normal."
--Albert Camus

"Some people wanted champagne and caviar when they should have had beer and hot dogs."
--Dwight D. Eisenhower

"Everybody has a heart. Except some people."
--Bette Davis

"Some people go to priests; others to poetry; I to my friends."
--Virginia Woolf

"Getting ahead in a difficult profession requires avid faith in yourself. That is why some people with mediocre talent, but with great inner drive, go so much further than people with vastly superior talent."
--Sophia Loren

"Some people will never learn anything, for this reason, because they understand everything too soon."
--Alexander Pope

"Some people are willing to pay the price and it's the same with staying healthy or eating healthy. There's some discipline involved. There's some sacrifices."
--Mike Ditka

"Some people bear three kinds of trouble – the ones they've had, the ones they have, and the ones they expect to have."
--H. G. Wells

"Some people approach every problem with an open mouth."
--Adlai E. Stevenson

"Some people regard discipline as a chore. For me, it is a kind of order that sets me free to fly."
--Julie Andrews

"It is not always by plugging away at a difficulty and sticking to it that one overcomes it; often it is by working on the one next to it. Some things and some people have to be approached obliquely, at an angle."
--Andre Gide

"Some people are born mediocre, some people achieve mediocrity, and some people have mediocrity thrust upon them."
--Joseph Heller

"Writing is not easy; some people want to write books but just can't put a story together."
--Jackie Collins

"Some people change when they think they're a star or something."
--Paris Hilton

"There are some people who have the quality of richness and joy in them and they communicate it to everything they touch. It is first of all a physical quality; then it is a quality of the spirit."
--Tom Wolfe

"Some people have been kind enough to call me a fine artist. I've always called myself an illustrator. I'm not sure what the difference is. All I know is that whatever type of work I do, I try to give it my very best."
--Norman Rockwell

"Really big people are, above everything else, courteous, considerate and generous – not just to some people in some circumstances – but to everyone all the time."
--Thomas J. Watson

Personalities

While teaching customer relations courses over the years, I began to see the value of assessing individual customer personalities. Most of us, at one time or another, have taken the Myers-Briggs Personality Test. This is a pretty good tool to discover who you are, but it's not very practical to reveal useful insights into customer behaviors. The personality tool I've found to be most effective in quickly determining how your customer might behave and how you can adjust in selling to them is DISC, a psychological inventory developed by John Geier and based on the work of psychologist William Moulton Marston.

Like Myers-Brigs, DISC is a preference inventory test, but it simplifies the results into four basic personality profiles:

Dominance - relating to control, power and assertiveness. People who score high in the "D" styles factor are very active in dealing with problems and challenges and are described as demanding, forceful, egocentric, strong willed, driving, determined, ambitious, aggressive, and pioneering.

Influence - relating to social situations and communication. People with high "I" scores influence others through talking and activity and tend to be emotional. They are described as convincing, magnetic, political, enthusiastic, persuasive, warm, demonstrative, trusting, and optimistic.

Steadiness - relating to patience, persistence, and thoughtfulness. People with high "S" styles scores want a steady pace, security, and do not like sudden change. High "S" individuals are calm, relaxed, patient, possessive, predictable, deliberate, stable, consistent, and tend to be unemotional and poker faced.

Conscientiousness - relating to structure and organization. People with high "C" styles adhere to rules, regulations, and structure. They like to do quality work and do it right the first time. High "C" people are careful, cautious, exacting, neat, systematic, diplomatic, accurate, and tactful.

Interesting, isn't it? But since we can't ask our customers to take the DISC, how can we use the system? Because the determination of the four personality types is based on only two variables – empathy and ego drive – you can size up people simply by asking yourself two questions: (1) Are they task oriented or people oriented? and (2) Are they proactive or responsive? D people are proactive and task oriented. I people are proactive and people oriented. S people are responsive and people oriented. And C people are responsive and task oriented.

It's important to note that there are neither good profiles nor bad profiles, and some people have mixtures. We all have value and can play important roles in business and life, even though we behave differently. For example (and speaking very generally), it's not surprising that in aerospace, defense, and technology businesses there are a lot of D program managers and a lot of I people in business development. S people hold organizations together and C people keep us on task. Friction occurs on the diagonals: D people are impatient with S people and S people don't like being bossed by D's. I people have trouble understanding why C people bury themselves in the details and C people prefer data over all that talk coming from I's.

How do you sell to the four personality types? Give D's two or three options so they can make the decision, not you. Give I's a vision so they can see how it all fits together (and how they fit into it all). Allow S's to collaborate on the solution and ease them into the final buy decision with examples of other satisfied customers. Provide C's with hard data and let them see for themselves the logic of your solution.

There is much, much more that can be done with DISC by taking the test and studying the results. But the basics should at least give you a sensitivity to the importance of personality differences in your customers and ways in which you can adjust to and perhaps benefit from those differences.

Personality Quotes

"An individual's self-concept is the core of his personality. It affects every aspect of human behavior: the ability to learn, the capacity to grow and change. A strong, positive self-image is the best possible preparation for success in life."
--Dr. Joyce Brothers

"If it weren't for caffeine I'd have no personality whatsoever!"
--Anonymous

"I am a deeply superficial person."
--Andy Warhol

"Personality is the glitter that sends your little gleam across the footlights and the orchestra pit into that big black space where the audience is."
--Mae West

"We know what a person thinks not when he tells us what he thinks, but by his actions."
--Isaac Bashevis Singer

"Talents are best nurtured in solitude: character is best formed in the stormy billows of the world."
--Johann Wolfgang von Goethe

"We should take care not to make the intellect our god; it has, of course, powerful muscles, but no personality."
--Albert Einstein

"Humility is no substitute for a good personality."
--Fran Lebowitz

"Don't compromise yourself. You are all you've got."
--Janis Joplin

"If I try to be like him, who will be like me?"
--Yiddish proverb

"He has not a single redeeming defect."
--Benjamin Disraeli

"The first time you meet Winston [Churchill] you see all his faults and the rest of your life you spend in discovering his virtues."
--Lady Constance Lytton

"Everyone is a moon and has a dark side which he never shows to anybody."
--Mark Twain

"I'd rather be strongly wrong than weakly right."
--Tallulah Bankhead

"Don't be so humble – you are not that great."
--Golda Meir

"I have an existential map; it has 'you are here' written all over it."
--Steven Wright

"The fellow that agrees with everything you say is either a fool or he is getting ready to skin you."
--Kin Hubbard

"Become who you are."
--Friedrich Nietzsche

"Always be yourself, express yourself, have faith in yourself, do not go out and look for a successful personality and duplicate it."
--Bruce Lee

"Personality is an unbroken series of successful gestures."
--F. Scott Fitzgerald

"Personality is immediately apparent, from birth, and I don't think it really changes."
--Meryl Streep

"Dogs got personality. Personality goes a long way."
--Quentin Tarantino

"Let your countenance be pleasant, but in serious matters somewhat grave."
--George Washington

"The most important single ingredient in the formula of success is knowing how to get along with people."
--Theodore Roosevelt

"Like an unchecked cancer, hate corrodes the personality and eats away its vital unity."
--Martin Luther King Jr.

"In the progress of personality, first comes a declaration of independence, then a recognition of interdependence."
--Henry Van Dyke

"One's personality can be understood from the people they mingle with."
--Kazi Shams

"Show me an actress who isn't a personality and I'll show you a woman who isn't a star."
--Katharine Hepburn

"Personality is to a man what perfume is to a flower."
--Charles Schwab

"It is far more impressive when others discover your good qualities without your help."
--Miss Manners (Judith Martin)

"It is the dull man who is always sure, and the sure man who is always dull."
--H. L. Mencken

Sales Process

Your company has all manner of processes. You have program processes to pull people and the global supply chain together for the benefit of customers and you have best practice processes to determine where and how you must bid to beat the competition and still make a profit. There are processes all around you. But have you ever actually used a sales process?

The process of selling goes so far back in mankind's memory that it seems to be something not even worth mentioning. It's understood. There are sellers (business development) and buyers (acquisition officials) and they meet and exchange goods and services for money. But if selling is so simple, why don't you close every sale? Have you ever been on a capture program and did all the right things and followed all the internal procedures and yet didn't make the sale? There are a lot of variables that can preclude deals, such as high-level politics, downward budget pressures and changing technical requirements. But maybe the customer didn't select your company because you weren't aware of sales fundamentals. Maybe you just didn't know how to close the deal.

It's helpful to think of sales as science based on behavioral research that can be distilled down to certain principles. All sales processes basically have the same essential elements. After you've gained thorough knowledge of the features and benefits of your product and service and you've identified and defined your target market and adjusted for the competition, you:

1. Prospect
2. Interview
3. Analyze Needs
4. Present
5. Negotiate
6. Close
7. Service and Follow-up

Do you go straight to #4 and never seem to get to #6? Why do you think that is? Is it because you believe your products and services are so superior that they'll sell themselves? It's a tempting thought, considering the eye-watering technologies your company can offer. But since people buy from people, perhaps better behaviors need to come into play to win more sales. Here's a set of behaviors and actions you might consider during the three phases of the sales process:

Before the Sale
- Be positively expectant
- Qualify the opportunity
- Identify a coach
- Research the customer's interests
- Set up the first calls and meetings

During the Sale
- Ask questions
- Build rapport
- Develop credibility

- Discover customer value
- Establish trust
- Provide a solution
- Address objections
- Confirm understanding
- Affirm decisions

After the Sale
- Extend the relationship
- Evaluate feedback
- Plan next action steps
- Seek additional opportunities
- Ask for referrals

So, what's in your process?

Process Quotes

"The process of scientific discovery is, in effect, a continual flight from wonder."
--Albert Einstein

"The system is that there is no system. That doesn't mean we don't have process. Apple is a very disciplined company, and we have great processes. But that's not what it's about. Process makes you more efficient."
--Steve Jobs

"Mothers all want their sons to grow up to be president, but they don't want them to become politicians in the process."
--John F. Kennedy

"When one has finished building one's house, one suddenly realizes that in the process one has learned something that one really needed to know in the worst way – before one began."
--Friedrich Nietzsche

"We do not learn; and what we call learning is only a process of recollection."
--Plato

"You can't process me with a normal brain."
--Charlie Sheen

"I know that two and two make four – and should be glad to prove it too if I could – though I must say if by any sort of process I could convert two and two into five it would give me much greater pleasure."
--Lord Byron

"The most difficult thing is the decision to act, the rest is merely tenacity. The fears are paper tigers. You can do anything you decide to do. You can act to change and control your life; and the procedure, the process is its own reward."
--Amelia Earhart

"The older I get the more wisdom I find in the ancient rule of taking first things first: a process which often reduces the most complex human problem to a manageable proportion."
--Dwight D. Eisenhower

"Life is a lively process of becoming."
--Douglas MacArthur

"If you can't describe what you are doing as a process, you don't know what you're doing."
--W. Edwards Deming

"Mind is the great lever of all things; human thought is the process by which human ends are ultimately answered."
--Daniel Webster

"The process of gathering knowledge does not lead to knowing."
--John Steinbeck

"And the thoughts of men are widened with the process of the suns; Knowledge comes, but wisdom lingers."
--Lord Alfred Tennyson

"A process cannot be understood by stopping it. Understanding must move with the flow of the process, must join it and flow with it."
--Frank Herbert

"Democracy is the process by which people choose the man who'll get the blame."
--Bertrand Russell

"Everyone is born a genius, but the process of living de-geniuses them."
--R. Buckminster Fuller

"I think some parents now look at a youngster failing as the final thing. It's a process, and failure is part of the process."
--Mike Krzyzewski

"Culture is the process by which a person becomes all that they were created capable of being."
--Thomas Carlyle

"After all these years, I am still involved in the process of self-discovery. It's better to explore life and make mistakes than to play it safe. Mistakes are part of the dues one pays for a full life."
--Sophia Loren

"The process by which banks create money is so simple that the mind is repelled."
--John Kenneth Galbraith

"No one can get an education, for of necessity education is a continuing process."
--Louis L'Amour

"The flood of print has turned reading into a process of gulping rather than savoring."
--Raymond Chandler

"Mere access to the courthouse doors does not by itself assure a proper functioning of the adversary process."
--Thurgood Marshall

"Life is one long process of getting tired."
--Samuel Butler

"Above all, I craved to seize the whole essence, in the confines of one single photograph, of some situation that was in the process of unrolling

itself before my eyes."
--Henri Cartier-Bresson

"Few businessmen are capable of being in politics, they don't understand the democratic process... Democracy isn't a business."
--Malcolm Forbes

"I'm for as much transparency in the newsgathering process as possible."
--Anderson Cooper

"I think making mistakes and discovering them for yourself is of great value, but to have someone else to point out your mistakes is a shortcut of the process."
--Shelby Foote

Promises

You may recall managers or colleagues saying the best way to do your business was to "under promise" and "over deliver." That keeps you in the "best value" area where customers will pay a little more for a premium solution. You may also have heard statements from US and foreign government customers that the notion of best value is past and that price alone will now drive all acquisition decisions. Which assertion is true? Perhaps neither.

A public survey of federal procurement officials asked what they were looking for from bidders. By their responses, key government decision makers rank-ordered:

1. A reputation for on-time, on-budget, in-scope delivery (57 percent of respondents)
2. Quality of the proposed solution (55 percent)
3. Knowledge (31 percent)
4. Price (27 percent)
5. Program management (24 percent)
6. Innovation (22 percent)

Price was fourth. What does that mean? Well, it could mean that your customers are concerned about costs and prices, but are more worried about your performance and their need for the right solution. If those are the overriding customer worries, how much confidence can they have in your company if you "under promise?" Will they even give you a chance to "over deliver?"

You don't want to "over promise." That's no way to build long-term trust. But you may want to think a little more about what exactly you're promising your customers. Do you remember the classic definition of a value proposition? *A combination of promised experiences and price, all in comparison to the best competing alternative, in return for a business relationship.* The value proposition has specific measurable events that will happen in the customer's experience and these results have consequences of value to the customer. The value proposition is your win strategy translated into the customer's context and it covers not only what you'll deliver, but more important, how you'll deliver it.

A good value proposition is simply a promise of what the customer will experience if they pick you. But they won't pick you if you don't have a solid promise to deliver a quality program on-time, on-budget and in-scope. What are you promising your customers?

Promise Quotes

"We must not promise what we ought not, lest we be called on to perform what we cannot."
--Abraham Lincoln

"Rarely promise, but, if lawful, constantly perform."
--William Penn

"One must have a good memory to be able to keep the promises that one makes."
--Friedrich Nietzsche

"Promises are like crying babies in a theater; they should be carried out at once."
--Norman Vincent Peale.

"An ounce of performance is worth pounds of promises."
--Mae West

"Nothing weighs lighter than an empty promise."
--German Proverb

"Eggs and oaths are easily broken."
--Danish Proverb

"Promises make debt, and debt makes promises."
--Dutch Proverb

"In the midst of great joy, do not promise anyone anything. In the midst of great anger, do not answer anyone's letter."
--Chinese Proverb

"A promise is a cloud; fulfillment is rain."
--Arabian Proverb

"The world did not make any promises."
--African Proverb

"In executing the duties of my present important station, I can promise nothing but purity of intentions, and, in carrying these into effect, fidelity and diligence."
--George Washington

"If we got one-tenth of what was promised to us in these acceptance speeches there wouldn't be any inducement to go to heaven."
--Will Rogers

"Politicians are the same all over. They promise to build bridges even when there are no rivers."
--Nikita Khrushchev

"It is useless to hold a person to anything he says while he is in love, drunk, or running for office."
--Shirley MacLaine

"Vote for the man who promises least; he'll be the least disappointing."
--Bernard M. Baruch

"Everyone is a millionaire where promises are concerned."
--Ovid

"Advertisements are now so numerous that they are very negligently perused, and it is therefore become necessary to gain attention by magnificence of promises and by eloquence sometimes sublime and

sometimes pathetic."
--Samuel Johnson

"Being of no power to make his wishes good: his promises fly so beyond his state, that what he speaks is all in debt; he owes for every word."
--William Shakespeare

"Hypocrisy can afford to be magnificent in its promises, for never intending to go beyond promise, it costs nothing."
--Edmund Burke

"Promises may fit the friends, but non-performance will turn them into enemies."
--Benjamin Franklin

"A promise must never be broken."
--Alexander Hamilton

"A guy will promise you the world and give you nothin', and that's the blues."
--Otis Rush

"Breach of promise is a base surrender of truth."
--Mohandas Gandhi

"Broken promises don't upset me. I just think, why did they believe me?"
--Jack Handy

"Those that are most slow in making a promise are the most faithful in the performance of it."
--Jean Jacques Rousseau

"The best way to keep one's word is not to give it."
--Napoleon Bonaparte

"It is not the oath that makes us believe the man, but the man the oath."
--Aeschylus

"The woods are lovely, dark and deep. But I have promises to keep, and miles to go before I sleep."
--Robert Frost

WHY CUSTOMERS QUIT - QUESTIONS

Why Customers Quit

In their "Reconnecting with Your Customer: Delivering Premium Customer Service" report, Frost & Sullivan analysts noted that "most businesses spend more on attracting new clients than on retention of the 'customer base,' and most lose half of their customer base every five years." That's pretty sobering, considering conventional wisdom that it costs about five times as much to get new customers as it does to fix and keep them.

Why do customers quit? According to the report, which included information from surveys of all kinds of companies across all business sectors:

- Only three percent move away (normal slippage)
- Five percent develop other relationships (they have lunch with you to say they are seeing someone else)
- Nine percent go to your competition (perhaps lured by a lower price)
- 14 percent are dissatisfied with your product (technical complications)
- But a whopping 68 percent leave because they sense an attitude of indifference toward them, demonstrated by the owner, manager, or an employee – or all of them.

That's stunning, isn't it? A friend working with an industry association reported similar results for declining membership in regional chapters and I've witnessed the same for volunteer groups in my community. Dr. Tom Barrett, with whom I've worked in the customer relations arena for many years, says every customer wants to know: Do you care? Do I trust you? Do you have anything to say? The caring comes before the sharing.

Doug Swimm, another colleague, lives up near Philadelphia. He says he drives across town to a particular pizza place, not because the pizza there is that much better than the pizza sold around the corner from his house, but because when he walks in they smile and acknowledge him and make him feel at home. They care. You perhaps recall being in a retail store looking for something in particular. The aggressively hovering salesperson is as annoying as the yawning salesperson behind the counter who doesn't seriously or enthusiastically try to answer your questions. Neither one appears to care about you.

Because we've all experienced these behaviors in the commercial world, they can echo into the federal marketplace. Government buyers don't want to be either pressured or ignored. They want to be helped. They want someone who cares more about the program mission than the profit margin on the contract. They want someone who cares and shows it. A senior military acquisition officer presenting to a group of defense industry representatives once put up this slide:

- Give us what we *want*…
- *When* you said you'd get it to us.
- Do it right the *first* time…
- *On* time.

You can't ignore the human factors involved in every government program, no matter if it's a relatively small Indefinite Delivery Indefinite Quantity (IDIQ) service contract or a several-billion-dollar multi-year platform franchise.

Customer satisfaction comes at a price. The finance people in your companies may be reluctant to part with funds for "keep sold" efforts until there are alarming indications of customer dissatisfaction (by which time it may be too late to do anything about the problem). So give your CFO some ROI figures to help justify the release of customer service money. According to the Strategic Planning Institute, businesses rated low on customer service averaged one percent ROS (Return on Sales, or profit margin) and lost market share at the rate of two percent per year. Companies that scored high in service averaged <u>12</u> percent ROS, <u>gained</u> market share at the rate of <u>six percent per</u> year and *charged higher prices*.

Another thing you can do is encourage a service-oriented atmosphere and give your employees the power to fix customer problems. Ritz Carleton is known for caring about its customers. The motto of their service personnel is, "We are Ladies and Gentlemen serving Ladies and Gentlemen." Their customer-facing employees are pre-authorized $2,000 per guest per day in customer service allowance funds to fix problems. That gives them the confidence and poise to make things right quickly (and without spending money every time). When you face a customer issue, are you so tightly controlled by your bosses that you cannot say anything other than, "We'll get back to you," knowing that it will take your contracts, legal, and communications staff weeks to craft a response? While they devise the perfect mitigation plan limiting your company's exposure, relations with your customer may go from bad to worse.

Perceptions are critically important. If your customer trusts you, believes you care, and sees indications that you are working hard to make the program a success, your customer will likely shift gaze to long-term concerns. But if your customer doesn't sense trust, doesn't feel that you care, and perceives no evidence that you're committed to shared program goals, every small issue becomes a big issue. Customer satisfaction comes at a price. A lot of it is paid by how you show support and by what you say and do every day.

Good customer satisfaction quotes are hard to find. Here are some blended customer satisfaction and personal satisfaction quotes. After all, if you satisfy your customer, you'll satisfy yourself.

Satisfaction Quotes

"If you work just for money, you'll never make it, but if you love what you're doing and you always put the customer first, success will be yours."
--Ray Kroc

"Unless you have 100% customer satisfaction, you must improve."
--Horst Schulz

"Choose to deliver amazing service to your customers. You'll stand out because they don't get it anywhere else."
--Kevin Stirtz

"Only a life lived in the service to others is worth living."
--Albert Einstein

"Good customer service costs less than bad customer service."
--Sally Gronow

"Always do more than is required of you."
--George Patton

"When the customer comes first, the customer will last."
--Robert Half

"We see our customers as invited guests to a party, and we are the hosts. It's our job every day to make every important aspect of the customer experience a little bit better."
--Jeff Bezos

"The first step in exceeding your customer's expectations is to know those expectations."
--Roy H. Williams

"Revolve your world around the customer and more customers will revolve around you."
--Heather Williams

"It is not fair to ask of others what you are not willing to do yourself."
--Eleanor Roosevelt

"Every company's greatest assets are its customers, because without customers there is no company."
--Michael LeBoeuf

"The best way to find yourself is to lose yourself in the service of others."
--Mahatma Gandhi

"Disciplining yourself to do what you know is right and important, although difficult, is the highroad to pride, self-esteem, and personal satisfaction."
--Margaret Thatcher

"I suppose if you've never bitten your nails, there isn't any way to explain the habit. It's not enjoyable, really, but there is a certain satisfaction – pride in a job well done."
--Anderson Cooper

"A man cannot be comfortable without his own approval."
--Mark Twain

"There are some days when I think I'm going to die from an overdose of satisfaction."
--Salvador Dali

"Whoever renders service to many puts himself in line for greatness – great wealth, great return, great satisfaction, great reputation, and great joy."
--Jim Rohn

"I get satisfaction of three kinds. One is creating something, one is being paid for it and one is the feeling that I haven't just been sitting on my ass all afternoon."
--William F. Buckley, Jr.

"I can't get no satisfaction."
--Mick Jagger

"There is no greater challenge than to have someone relying upon you; no greater satisfaction than to vindicate his expectation."
--Kingman Brewster

"There's always a little bit of personal satisfaction when you prove somebody wrong."
--Drew Brees

"The ultimate victory in competition is derived from the inner satisfaction of knowing that you have done your best and that you have gotten the most out of what you had to give."
--Howard Cosell

"Action itself, so long as I am convinced that it is right action, gives me satisfaction."
--Jawaharlal Nehru

"All the satisfaction I need... comes when I step out onstage and see the people."
--Harry Connick, Jr.

"I get satisfaction out of seeing stuff that makes real change in the real world. We need a lot more of that and a lot less abstract stuff."
--Temple Grandin

"There is a great satisfaction in building good tools for other people to use."
--Freeman Dyson

"If a man has talent and can't use it, he's failed. If he uses only half of it, he has partly failed. If he uses the whole of it, he has succeeded, and won a satisfaction and triumph few men ever know."
--Thomas Wolfe

"As far as playing jazz, no other art form, other than conversation, can give the satisfaction of spontaneous interaction."
--Stan Getz

"Success is finding satisfaction in giving a little more than you take."
--Christopher Reeve

"The bargain that yields mutual satisfaction is the only one that is apt to be repeated."
--B. C. Forbes

Questions

What do Socrates, John Marshall, and Colombo the TV detective have in common? All three knew the power of questions. After he finished writing *To Sell is Human: The Surprising Truth about Moving Others*, Daniel H. Pink said that he wished he had expanded the section on questions. The more he examined the use of questions to understand other people and help guide them to make good decisions, the more he began to understand their power. Once again, a glaring truth about the human condition that we ignore until someone reminds us of it. My colleague Dr. Tom Barrett, who had a counseling practice for many years, knows how sincere and artful questions at appropriate moments can elicit powerful responses. Can you recall when a good friend or loved one asked one or two simple, heart-felt questions at an appropriate moment and cracked you open like a walnut?

Good questions can elicit positive responses. In the late 1990s I was the Defense Attaché at the American Embassy in Prague. I also was heading up the Security Assistance Office, which was supporting the Czech entry into NATO. I was continuously visited by US defense contractors seeking my personal support in selling their wares. Many of them were leaning on me to gain access to key Czech decision makers and pushing me to favor their products over the competition. This was difficult for me, because the Czechs had a limited budget to fund all the requirements setting up interoperability with NATO countries and I felt my role was to help the Czechs and the Americans equally. One contractor visited my office during a busy time and started in on his pitch. I was distracted by all the things I had to do that day and I was only half-listening. Then he asked, "What do you think would be the best outcome for the Czechs in this area?" In a flash moment he turned me from a disinterested bystander into an eager advocate for his program.

Poor questions elicit unimportant information. Many customer relations gurus tout the mantra "Listen to the Customer." But isn't the quality of what you are listening to directly proportional to the quality of your questions? A while back, my friend Joe Verderame and I were at a business development conference in Florida. It was a beautiful evening by the hotel pool and we were losing track of our margaritas, but since we were staying at the hotel we were relaxed about it. Joe is a great business developer. In our increasingly philosophical conversation we turned to the subject of whether it was better to hire business developers from the customer base or former program managers from inside the company. Joe was emphatic that it was better to hire from within, since that's where he came from. I was equally insistent that you can only truly understand a government customer if you've been one, like me. Joe retorted, "I can outsell you Bubbas any day. Your hand-holding and door-opening aren't nearly as powerful as my Smart Questions." As the evening wore on, I became convinced that he was on to something.

You, too, can "Beat the Bubbas" with Smart Questions leading to more intelligent conversations. According to Joe, government customers really want solutions, not friends. They want substantive interactions with well-informed company representatives. Therefore, use your knowledge of programs (especially lessons-learned details) to engage customers with questions that demonstrate not only your technical knowledge, but your understanding of their problems. One smart question delivered at a strategic moment can elevate you above your competitors and lead to continued engagements.

In subsequent encounters with your customer, rely on information brokering. If you provide your customers additional knowledge that they can consider and perhaps act on, they may be inclined to provide you new information that can lead to greater understanding not only of their technical requirements, but their solution desires. Continued interaction in information sharing will allow both you and your customer to shape a win-win RFP.

As you prepare your questions for your customer encounter (you do prepare beforehand, don't you?) take some advice from Arlene Johnson of Sinequanon Group in Dallas. Arlene runs Customer Value Conversation sessions and conducts Miller Heiman sales courses adapted for companies doing business with government customers. She is also the author of *Success Mapping*, a roadmap tool to help you get to where you want to go in your career. Arlene suggests that before the sales call, ask yourself:

- What might be the value to the customer to tell us what we need?
- What three open-ended questions, if answered, would help us qualify or differentiate?
- What customer recommendation would help favorably position us and move the decision cycle forward?
- What would be a "best action" commitment we could ask the customer to take?

I love this list, especially the recommendation for three open-ended questions. Open-ended questions allow customers to talk and you to listen. Artfully applied, they can encourage customers to take pauses in their busy schedules to think beyond the day's headlines and give you clues to opportunities. Save your closed questions to summarize the meeting or move your customer to closing.

Unfortunately, many sales meetings go bad either by failure to prepare and coordinate beforehand or over-scripting to the point that the meeting becomes an interrogation, not a conversation. One of the greatest interviewers, Larry King, was successful because he combined thorough research with spontaneity. He made his guests so relaxed and comfortable that they revealed deeply personal things to millions of people – and felt good about it.

So what questions will you ask next time?

Question Quotes

"A question that sometimes drives me hazy: am I or are the others crazy?"
--Albert Einstein

"We are all agreed that your theory is crazy. The question which divides us is whether it is crazy enough to have a chance of being correct. My own feeling is that it is not crazy enough."
--Niels Bohr

"A wise man can learn more from a foolish question than a fool can learn from a wise answer."
--Bruce Lee

"A prudent question is one-half of wisdom."
--Francis Bacon

"I wish I had an answer to that because I'm tired of answering that question."
--Yogi Berra

"It is not enough to be busy. So are the ants. The question is: What are we busy about?"
--Henry David Thoreau

"Question everything. Learn something. Answer nothing."
--Euripides

"To be, or not to be, that is the question."
--William Shakespeare

"Life is an unanswered question, but let's still believe in the dignity and importance of the question."
--Tennessee Williams

"In all affairs it's a healthy thing now and then to hang a question mark on the things you have long taken for granted."
--Bertrand Russell

"Always the beautiful answer who asks a more beautiful question."
--e. e. cummings

"I'd asked around 10 or 15 people for suggestions. Finally one lady friend asked the right question, 'Well, what do you love most?' That's how I started painting money."
--Andy Warhol

"A man may fulfill the object of his existence by asking a question he cannot answer, and attempting a task he cannot achieve."
--Oliver Wendell Holmes

"No question is so difficult to answer as that to which the answer is obvious."
--George Bernard Shaw

"A sudden bold and unexpected question doth many times surprise a man and lay him open."
--Francis Bacon

"The stupidity of people comes from having an answer for everything. The wisdom of the novel comes from having a question for everything."
--Milan Kundera

"He must be very ignorant for he answers every question he is asked."
--Voltaire

"There is hardly a political question in the United States which does not sooner or later turn into a judicial one."
--Alexis de Tocqueville

"Never ask a bore a question."
--Mason Cooley

"The power to question is the basis of all human progress."
--Indira Gandhi

"It is not the answer that enlightens, but the question."
--Eugene Ionesco

"To spell out the obvious is often to call it in question."
--Eric Hoffer

"The common question that gets asked in business is, 'why?' That's a good question, but an equally valid question is, 'why not?' "
--Jeff Bezos

"For your information, I would like to ask a question."
--Samuel Goldwyn

"It is not every question that deserves an answer."
--Publilius Syrus

"If you do not know how to ask the right question, you discover nothing."
--W. Edwards Deming

"I asked Dalai Lama the most important question that I think you could ask – if he had ever seen Caddyshack."
--Jesse Ventura

"He who cannot eat horsemeat need not do so. Let him eat pork. But he who cannot eat pork, let him eat horsemeat. It's simply a question of taste."
--Nikita Khrushchev

"A politician is a statesman who approaches every question with an open mouth."
--Adlai E. Stevenson

"Charm is a way of getting the answer yes without asking a clear question."
--Albert Camus

"To every answer you can find a new question."
-- Yiddish Proverb

CUSTOMER RELATIONS - RESOLUTIONS - RISK

Customer Relations

If you're looking to improve your business unit's customer relations and don't know where to start, consider taking clues from other companies, company leaders and your customers. Here are examples of each.

<u>Another Company</u>

In the first few pages of his book, *Winning*, Jack Welch discusses corporate values and behaviors and literally gushes over Bank One's list for treating customers well. They are worth a hard look:

- Never let profit center conflicts get in the way of doing what is right for the customer.
- Give customers a good, fair deal. Great customer relationships take time. Do not try to maximize short-term profits at the expense of building those enduring relationships.
- Always look for ways to make it easier to do business with us.
- Communicate daily with your customers. If they are talking to you, they can't be talking to a competitor.
- Don't forget to say thank you.

<u>A Company Leader</u>

I was in a meeting being addressed by Bob Stevens, the former CEO of Lockheed Martin. I had the pleasure to work for him and admired him greatly. Most people who heard Bob speak came away with thoughtful insights. He spoke of difficult subjects, outlined how he expected his business leaders to meet those challenges and then opened it up to the audience. There followed a series of questions on difficult and complicated customer relationships. He gave this response: "We are a company with big business resources, but we need small business customer relationships." That still resonates. Even if your company is a large defense contractor, your customers want and need to have the feeling that you are as easy to do business with as the proprietor of a small corner store.

A Government Customer

The subject of customer relations is always on my mind. Once I was in a waiting room of the Walter Reed National Medical Center, wondering what our government customers would say about customer relations, if they were the ones providing products and services to others. I looked up and on the wall was a sign titled "Standards of Customer Service Excellence." It wasn't a fancy mass-produced sign, merely a simple list prepared on a computer and printed out on a laser printer. It had the appearance of local customer relations values and I have to say that the clinic employees did indeed give me a positive customer experience. Here is what the sign said:

- Treat everyone with courtesy, compassion and respect.
- Ensure the privacy, confidentiality and dignity of others.
- Be professional in appearance and behavior.
- Ensure that workspaces are neat, clean, safe and quiet.
- Be an effective communicator.
- Take ownership of problems and be a problem-solver.
- Seek to understand and meet our customers' needs.
- Be reliable and trustworthy.
- Be a team player.

These are good words for anybody. That they've come from a government service provider strongly suggests other government customers would also value these behaviors coming from you.

Customer Relations Quotes

"The easiest kind of relationship for me is with ten thousand people. The hardest is with one."
--Joan Baez

"Tell me, I'll forget. Show me, I'll remember. Involve me, I'll understand."
--Chinese proverb

"I've learned that people will forget what you said, people will forget what you did, but people will never forget how you made them feel."
--Maya Angelou,

"If you had to identify, in one word, the reason why the human race has not achieved, and never will achieve, its full potential, that word would be 'meetings'."
--Dave Barry

"Always remember that you are absolutely unique. Just like everyone else."
--Margaret Mead

"All lasting business is built on friendship."
--Alfred A. Montapert

"How far you go in life depends on your being tender with the young, compassionate with the aged, sympathetic with the striving and tolerant of the weak and strong. Because someday in your life you will have been all of these."
--George Washington Carver

"Greater profit in business comes from repeat customers, customers that boast about your project or service, and that bring friends with them."
--W. Edwards Deming

"Spend a lot of time talking to customers face to face. You'd be amazed how many companies don't listen to their customers."
--Ross Perot

"More business is lost every year through neglect than through any other cause."
--Rose Kennedy

"When you lose a customer, it can be tempting to tell each other, 'That customer's not very sharp. They just made the wrong decision'."
--Bill Gates

"Nothing gives a person so much advantage over another as to remain always cool and unruffled under all circumstances."
--Thomas Jefferson

"A business absolutely devoted to service will have only one worry about profits. They will be embarrassingly large."
--Henry Ford

"Statistics suggest that when customers complain, business owners and managers ought to get excited about it. The complaining customer represents a huge opportunity for more business."
--Zig Ziglar

"If a man loves the labour of his trade, apart from any question of success or fame, the gods have called him."
--Robert Louis Stevenson

"The stupid neither forgive nor forget; the naive forgive and forget; the wise forgive but do not forget."
--Thomas Szasz

"Be who you are and say what you feel, because those who mind don't matter and those who matter don't mind."
--Theodor Seuss Geisel, aka Dr. Seuss

"Our prime purpose in this life is to help others. And if you can't help them, at least don't hurt them."
--Tenzin Gyatso, the 14th Dalai Lama

"A smile is the chosen vehicle for all ambiguities."
--Herman Melville

"It takes less time to do a thing right than to explain why you did it wrong."
--Henry Wadsworth Longfellow

"Give the public everything you can give them, keep the place as clean as you can keep it, keep it friendly."
--Walt Disney

"If you would persuade, you must appeal to interest rather than intellect."
--Benjamin Franklin

"Reason often makes mistakes, but conscience never does."
--Josh Billings

"Anyone who takes himself too seriously always runs the risk of looking ridiculous; anyone who can consistently laugh at himself does not."
--Vaclav Havel

"It is absurd to divide people into good and bad. People are either charming or tedious."
--Oscar Wilde

"Conversation would be vastly improved by the constant use of four simple words: I do not know."
--Andre Maurois

"You will only be remembered for two things: the problems you solve or the ones you create."
--Mike Murdock

"You never really understand a person until you consider things from his point of view."
--Harper Lee

"Once the game is over, the King and the pawn go back in the same box."
--Italian Proverb

Risk

In troubled times responsible business people assess the market environment and take stock of where their programs are. Experienced business people understand that volatile markets include both risks and opportunities. But discussion of risks soon starts to crowd out discussion of opportunities. That can lead to navel-gazing paralysis of action in business development. Contributing to this are the various types of risk and what they mean to various business departments. Here are some samples:

- Technical Risk - changes that could impact the project or technologies that may not work as expected
- Legal Risk - the potential for adverse regulatory or civil actions
- Security Risk - events that could result in the compromise of organizational assets

- Financial Risk - the uncertainty of a return and the potential for financial loss

Business people often confuse the words "risk" and "uncertainty" and interchange them just like they do with the words "strategy" and "planning." In his 1921 book *Risk, Uncertainty and Profit* (still considered a classic), Frank Knight of the University of Chicago established a distinction between the two. Risk is randomness with knowable probabilities and uncertainty is randomness with unknowable probabilities. Risk is quantifiable either from prior knowledge or empirical observation and, most importantly, is measurable and can be avoided, mitigated and managed. Because risks are measurable, they are distinctly separate from the uncertainties you can neither define nor affect.

By separating measurable risk from uncertainty you can better determine outcomes and predict what might happen in the future. Take for example an activity you are familiar with – flying. When people first began flying they felt it was extremely risky and they were right. But engineers began to figure out what could be known and acted upon. Over time, technological advances were introduced, which made flying safer and dangers were reduced to such an extent that flying is now one of the least risky modes of transportation. Of course, there still are unpredictable dangers to flying, but until they can be defined and measured they will remain uncertainties.

Uncertainty breeds fear. Our human instinct is to rely on fear and hesitation to keep us out of danger. In a way, fear is a gift (see *The Gift of Fear* by Gavin de Becker). Fear helps us survive. But when real dangers aren't separated from irrational dangers we can actually make irrational choices and increase risk. Risk involves choice. Unlike uncertainty, measurable risk allows the possibility that a course of action (or inaction) will lead to a change in outcome. And here is where it becomes confusing again. The outcomes of our choices can be positive as well as negative. Risk has a duality, but most business people focus only on the negative.

Everyone wants to reduce or mitigate risk (terms also used interchangeably). When you are uncertain, you don't know what will happen next. But if you separate the risks from the uncertainties and engage a plan to reduce them, then you likely will reduce chances of unfavorable outcomes. But what if you increase risk? Can't you also increase the chances for a favorable outcome? Yes. While risk mitigation reduces the probability/impact of unfortunate events, purposeful risk management can also permit you to make choices toward maximizing opportunities.

So now you may be asking, "What does this have to do with customers?" Everything. Customers, like us, are prone to irrational fears in times of perceived uncertainty. Key decision makers, buyers, and users don't like this and the more stress they have in their environments, the more they want to avoid it. Even when they are able to separate real risks from uncertainties, their default setting on program decisions in troubled times is "no choice." Regrettably, this tendency reduces their probability of success and reduces your probability of contracts.

To reduce the fear of risk in the minds of your customers, you can do at least three things. The first is to continue to identify/avoid/mitigate/manage measurable internal risks that are actual threats to your corporate well-being. The second is to identify and increase calculated risks to create and improve programs for the benefit of your customers. The third is to make sure you explain the second thing to the customer. If a customer knows you are taking risks on their behalf, it signals you are all in it together. If you want to decrease customer risk, you need to consider increasing your own risk. Risk can be opportunity.

Risk Quotes

"It seems to be a law of nature, inflexible and inexorable, that those who will not risk cannot win."
--John Paul Jones

"Only those who will risk going too far can possibly find out how far one can go."
--T. S. Eliot

"Prophesy is a good line of business, but it is full of risks."
--Mark Twain

"This report, by its very length, defends itself against the risk of being read."
--Winston Churchill

"Hesitation increases in relation to risk in equal proportion to age."
--Ernest Hemingway

"Risk comes from not knowing what you're doing."
--Warren Buffett

"Living at risk is jumping off the cliff and building your wings on the way down."
--Ray Bradbury

"Be wary of the man who urges an action in which he himself incurs no risk."
--Seneca

"It is part of a good man to do great and noble deeds, though he risk everything."
--Plutarch

"The policy of being too cautious is the greatest risk of all."
--Jawaharlal Nehru

"Anyone who takes himself too seriously always runs the risk of looking ridiculous; anyone who can consistently laugh at himself does not."
--Vaclav Havel

"The only risk of failure is promotion."
--Scott Adams

"There are one hundred men seeking security to one able man who is willing to risk his fortune."
--J. Paul Getty

"Only a few act – the rest of us reap the benefits of their risk."
--Wynton Marsalis

"The risk of a wrong decision is preferable to the terror of indecision."
--Maimonides

"If you're not a risk taker, you should get the hell out of business."
--Ray Kroc

"Security is mostly a superstition. Life is either a daring adventure or nothing."
--Helen Keller

"Go out on a limb. That's where the fruit is."
--Will Rogers

"I want to stay as close to the edge as I can without going over. Out on the edge you see all kinds of things you can't see from the center."
--Kurt Vonnegut

"Do one thing every day that scares you."
--Eleanor Roosevelt

"The dangers of life are infinite, and among them is safety."
--Goethe

"It was a bold person that first ate an oyster."
--Jonathan Swift

"Take calculated risks. That is quite different from being rash."
--General George Patton

"I can accept failure. Everybody fails at something. But I can't accept not trying. Fear is an illusion."
--Michael Jordan

"If it's a good idea, go ahead and do it. It's much easier to apologize than it is to get permission."
--Rear Admiral Grace Hopper

"The path is smooth that leadeth on to danger."
--Shakespeare

"Two roads diverge in a wood, and I took the one less traveled by, and that has made all the difference."
--Robert Frost

"When you come to a fork in the road, take it!"
--Yogi Berra

"Yes, risk taking is inherently failure-prone. Otherwise, it would be called sure-thing-taking."
--Jim McMahon

"Don't let the fear of striking out hold you back."
--George Herman (Babe) Ruth, Jr.

"First ponder, then dare."
--Helmuth Johannes Ludwig, Graf von Moltke

"Stupid risks make life worth living."
--Homer Simpson

"He who risks and fails can be forgiven. He who never risks and never fails is a failure in his whole being."
--Paul Tillich

"Nothing will ever be attempted if all possible objections must first be overcome."
--Samuel Johnson

"There are costs and risks to a program of action, but they are far less than the long-range risks and costs of comfortable inaction."
--John F. Kennedy

"If things seem under control, you are just not going fast enough."
--Mario Andretti

S

SALES - SERVICE - STRATEGY AND TACTICS

Sales

Near the beginning of each customer relations class I teach, I usually ask, "Who's in sales?" A few hands go up. I wait. Then someone whispers the obvious, "We're all in sales." It doesn't matter where your job is along the business continuum – prospecting, selling, or keeping customers – you're all involved in sales.

The trick is that your government customers don't want you to be in "sales." That's why your front-end salespeople use the term "business development" (or market development) on their calling cards. In fact, your government customers don't really see themselves as customers, either. They prefer to be called "acquisition" officials. Words are important, but let's not kid ourselves. You're selling and, if you do your job properly, they're buying.

As noted in an earlier section, most people in business development got their job one of two ways. They either came up through the programs or transferred in from the customer community. Both groups are familiar with the technical details of their company's offerings and the unique characteristics of their customer sets. But many in both groups have only scant knowledge of sales techniques. Some business development professionals are very good at their jobs due to their engaging personalities, but they have difficulty explaining to others why they are successful and cannot seem to pass on their skill sets to others.

To some of your business development colleagues, the loss of a sale may invoke the perception that it was the fault of the customer. It's true that government customers have great difficulties making buying decisions, but it's possible some of those difficulties could be eased if you employed better techniques to help them in those decisions. The greatest sales people are those who don't really appear to be selling at all; they put all their efforts into helping their customers make selections. Can you do that?

Many commercial companies that are successful in sales growth use personality assessments to screen sales force applicants. Some personalities are great at outside sales and some are better at inside sales. Some personalities are suited to sales management and some are tailor-made for customer service. All have value, but each excels in certain positions over others. Are you in the right sales job?

Everyone could benefit from a little sales training. Why? You can learn new skills, get customers faster, cross-sell better, gain more company revenues, earn higher pay, and enjoy your job more. On average, Fortune 500 companies give their sales staff 40 hours of sales training. And this doesn't include annual refreshers, product familiarization, and process improvement programs. How much sales training have you had?

Please think about how you and your team could benefit from understanding the dynamics of the sales process and explore the idea that maybe, just maybe, a little sales training could help. There are many good sales training courses available, but your business is somewhat different than the commercial product sector and off-the-shelf sales courses don't often seem suitable to what you do. You might consider courses taught by people who have worked in the aerospace, defense and technology sector, such as *Top Ten Selling Skills* by John Asher and *Customer Value Conversations* by Arlene Johnson (see their contact information in the back of this book).

Sales Quotes

"A sale is not something you pursue, it's what happens to you while you are immersed in serving your customer."
--Unknown

"Everyone lives by selling something."
--Robert Louis Stevenson

"Sales are contingent upon the attitude of the salesman – not the attitude of the prospect."
--W. Clement Stone

"The fact is, everyone is in sales. Whatever area you work in, you do have clients and you do need to sell."
--Jay Abraham

"The key is not to call the decision maker. The key is to have the decision maker call you."
--Jeffrey Gitomer

"I have never worked a day in my life without selling. If I believe in something, I sell it, and I sell it hard."
--Estée Lauder

"Catch a man a fish, and you can sell it to him. Teach a man to fish, and you ruin a wonderful business opportunity."
--Unknown

"Most people think 'selling' is the same as 'talking'. But the most effective salespeople know that listening is the most important part of their job."
--Roy Bartell

"The sale most often goes to the most interested party."
--Steve Chandler

"Timid salesmen have skinny kids."
--Zig Ziglar

"To succeed in sales, simply talk to lots of people every day. And here's what's exciting – there are lots of people!"
--Jim Rohn

"Some people fold after making one timid request. They quit too soon. Keep asking until you find the answers. In sales there are usually four or five no's before you get a yes."
--Jack Canfield

"Don't sell life insurance. Sell what life insurance can do."
--Ben Feldman

"Your most important sale is to sell yourself to yourself."
--Unknown

"The sale begins when the customer says yes."
--Harvey MacKay

"Obstacles are necessary for success because in selling, as in all careers of importance, victory comes only after many struggles and countless defeats."
--Og Mandino

"You've got to be success minded. You've got to feel that things are coming your way when you're out selling; otherwise, you won't be able to sell anything."
--Benjamin Jowett

"Don't Sell – Solve."
--Unknown

"The point to remember about selling things is that, as well as creating atmosphere and excitement around your products, you've got to know what you're selling."
--Stuart Wilde

"A smart salesperson listens to emotions not facts."
--Unknown

"In sales there are going to be times when you can't make everyone happy. Don't expect to and you won't be disappointed. Just do your best for each client in each situation as it arises. Then, learn from each situation how to do it better the next time."
--Tom Hopkins

"Confidence and enthusiasm are the greatest sales producers in any kind of economy."
--O. B. Smith

"A salesman, like the storage battery in your car, is constantly discharging energy. Unless he is recharged at frequent intervals he soon runs dry. This is one of the greatest responsibilities of sales leadership."
--R.H. Grant

"To satisfy our customers' needs, we'll give them what they want, not what we want to give them."
--Steve James

"Forget about the business outlook, be on the outlook for business."
--Paul J. Meyer

"Internalize the Golden Rule of sales that says: All things being equal, people will do business with, and refer business to, those people they know, like, and trust."
--Bob Burg

"I think selling techniques are basically the same in every country, except there are different cultures that have different methods of negotiating."
--George Ross

"Always be closing... That doesn't mean you're always closing the deal, but it does mean that you need to be always closing on the next step in the process."
--Shane Gibson

"People get caught up in wonderful, eye-catching pitches, but they don't do enough to close the deal. It's no good if you don't make the sale. Even if your foot is in the door or you bring someone into a conference room, you don't win the deal unless you actually get them to sign on the dotted line."
--Donald Trump

"You don't close a sale, you open a relationship if you want to build a long-term, successful enterprise."
--Patricia Fripp

Service

When people think about customer service they often think of call centers staffed by employees trained to work product problems. Customer support activities certainly are an important segment of customer service, but they're normally at the back end of the customer relationship. When you're looking for new business or in the process of capturing business, do you ever think about customer service? Shouldn't you?

Customers don't normally express their desire for customer service in conversations or requests for proposals, but it's there. Sometimes customer interest in good service is embedded in past performance ratings for new starts, sometimes in their reluctance to part from a contractor they feel good about even with the prospect of saving money by recompeting the contract.

In the aerospace and defense industry the differences in offerings by competitor companies often aren't all that great, especially in the eyes of the customers. With all things being equal, a customer service oriented company can gain competitive advantage. But what, exactly, is customer service? According to the business school gurus, customer service is a set of activities put in place to raise the level of satisfaction – the customer's feeling that a product or service will meet or has met expectations. Notice the word *feeling*.

In so many ways, customer service is more how you do something than what you do. It's in listening to the customer, avoiding surprises, paying attention to details, responding quickly to requests, and following up to make sure the customer feels the program is on track. In the customer's mind, feelings are facts. Being nice to customers is only a part of good customer service. You also have to put systems and people in place to center your business around the customer and do the job right the first time.

A lot of business leaders mistakenly believe that customer service is something somebody else needs to worry about. Actually, the only way to implement a customer service oriented business culture is top down. Customer service should be included one way or another in annual goals and performance ratings, and especially in the characters and personalities of the employees leaders hire and promote. Leaders set the customer service tone. If they don't motivate their employees toward serving customers and empower them to solve customer problems quickly, nothing will change. To be really effective, leaders should treat their employees like they want their employees to treat their customers.

How do you approach customers? In business development, are you just selling to customers or are you helping your customers find solutions that work? In program management, are you just fulfilling the terms of the contract or are you serving your customers and making sure that you're meeting their expectations every day?

Something to think about: good customer service is underappreciated in business – except by the customers.

Service Quotes

"If you take good care of the customers, they come back. If you take good care of the products, they don't come back."
--Stanley Marcus, Neiman-Marcus

"If you help enough people get what they want, you will get what you want."
--Zig Ziglar

"Do what you do so well that they will want to see it again and bring their friends."
--Walt Disney

"If Franz Kafka were alive today he'd be writing about customer service."
--Jonathan Alter

"The single most important thing to remember about any enterprise is that there are no results inside its walls. The result of a business is a satisfied customer."
--Peter Drucker

"There are no traffic jams along the extra mile."
--Roger Staubach

"One of the deep secrets of life is that all that is really worth doing is what we do for others."
--Lewis Carol

"The goal as a company is to have customer service that is not just the best, but legendary."
--Sam Walton

"The way to gain a good reputation, is to endeavor to be what you desire to appear."
--Socrates

"Business is not just doing deals; business is having great products, doing great engineering, and providing tremendous service to customers. Finally, business is a cobweb of human relationships."
--H. Ross Perot

"Be everywhere, do everything, and never fail to astonish the customer."
--Macy's Motto

"Biggest question: Isn't it really 'customer helping' rather than customer service? And wouldn't you deliver better service if you thought of it that way?"
--Jeffrey Gitomer

"Every great business is built on friendship."
--J.C. Penney

"Give what you have to somebody, it may be better than you think."
--Henry Wadsworth Longfellow

"Our business is about technology, yes. But it's also about operations and customer relationships."
--Michael Dell

"What do we live for if not to make life less difficult for each other?"
--George Eliot

"I don't know what your destiny will be, but one thing I know: the ones among you who will be really happy are those who have sought and found how to serve."
--Albert Schweitzer

"If you want to lift yourself up, lift up someone else."
--Booker T. Washington

"I'll invest my money in people."
--W.K. Kellogg

"If we do not lay out ourselves in the service of mankind whom should we serve?"
--John Adams

"If you don't genuinely like your customers, chances are they won't buy."
--Thomas Watson

"Right or wrong, the customer is always right."
--Marshall Field

"We can believe that we know where the world should go. But unless we're in touch with our customers, our model of the world can diverge from reality. There's no substitute for innovation, of course, but innovation is no substitute for being in touch, either."
--Steve Ballmer

"If you do build a great experience, customers tell each other about that. Word of mouth is very powerful."
--Jeff Bezos

"Dealing with people is probably the biggest problem you face, especially if you are in business. Yes, and that is also true if you are a housewife, architect or engineer."
--Dale Carnegie

"I think we're having fun. I think our customers really like our products. And we're always trying to do better."
--Steve Jobs

"The greatest discovery of my generation is that human beings can alter their lives by altering their attitude of mind."
--William James

"We don't want to push our ideas on to customers, we simply want to make what they want."
--Laura Ashley

"If you work just for money, you'll never make it, but if you love what you're doing and you always put the customer first, success will be yours."
--Ray Kroc

"Well done is better than well said."
--Benjamin Franklin

Strategy and Tactics

"Strategy without tactics is the slowest route to victory. Tactics without strategy is the noise before defeat." (Sun Tzu [500BC], Author of The Art of War)

How often do you hear the term "strategy" used in marketing decision meetings? Maybe too often. If you're former military like me you're accustomed to the idea that a strategy is a plan of action designed to achieve a particular goal, which is separated into specific objectives. Tactics, a term much less used in business circles, can be thought of as actions taken to achieve those objectives. In military thought, a strategy is big picture and longer term; a tactic is focused and immediate. I believe this model works well in business, too.

Some of the left-brain confusion over use of the term strategy may come from game theory, in which a particular strategy is only one of several options a player might choose. Right brain confusion over strategy could come from business school models where definitions for "vision" and "goals" are intermixed along with those for "objectives" and "tasks."

The important thing is to acknowledge that the term strategy can have multiple meanings and try to discern which definition is being used in which setting. Whatever you call them, both strategies and tactics are needed. Grand visions for gaining customers won't get you very far unless you set plans in motion. And day-to-day interactions with customers won't get you those big sales unless you have a thought-out strategy. Winning sales teams define (1) what end state is desired, (2) what strategy is best, (3) what tactical actions are needed, (4) who will do them and (5) by what date.

In my opinion, sports analogies are always appropriate in illustrating fundamental truths. Take football. The goal is to win in the allotted time. The offensive strategy is to gain yards and score more points than the opponent. Typical offensive strategies include the Option, West Coast, Spread, Smashmouth, etc. Offensive tactics are the individual plays, such as: Draw, Counter Run, Sweep, Screen Pass, Student Body Right, etc. Admittedly, this is a simple description of a complex system of systems (there also are defensive and special teams strategies), but you get the idea.

The question is, when you go to market with potential and existing customers, are you overly focused on lots of activities (everyone out for a pass) or employing both winning strategies and winning tactics (a balanced offense)? Here are some football quotes to help reinforce these concepts.

Football Quotes

"Setting a goal is not the main thing. It is deciding how you will go about achieving it and staying with that plan."
--Tom Landry

"If anything goes bad, I did it. If anything goes semi-good, we did it. If anything goes really good, then you did it. That's all it takes to get people to win football games for you."
--Paul "Bear" Bryant

"We didn't lose the game; we just ran out of time."
--Vince Lombardi

"The road to Easy Street goes through the sewer."
--John Madden

"One man practicing sportsmanship is better than a hundred teaching it."
--Knute Rockne

"Ability is what you're capable of doing. Motivation determines what you do. Attitude determines how well you do it."
--Lou Holtz

"Thanksgiving dinners take eighteen hours to prepare. They are consumed in twelve minutes. Half-times take twelve minutes. This is not coincidence."
--Erma Bombeck

"If you don't invest very much, then defeat doesn't hurt very much, and winning is not very important."
--Dick Vermeil

"You may not win them over, but if you hang around long enough, you'll wear them out."
--Phil Simms

"You don't win on emotion. You win on execution."
--Tony Dungy

"Stubbornness is a virtue if you're right. It's only a character flaw if you're wrong."
-- Chuck Noll

"Nobody in football should be called a genius. A genius is a guy like Norman Einstein."
--Joe Theismann

"Baseball players are smarter than football players. How often do you see a baseball team penalized for too many men on the field?"
--Jim Bouton

"Sure, luck means a lot in football. Not having a good quarterback is bad luck."
--Don Shula

"In life, as in a football game, the principle to follow is: Hit the line hard."
--Theodore Roosevelt

"What's the worst thing that can happen to a quarterback? He loses his confidence."
--Terry Bradshaw

"Confidence doesn't come out of nowhere. It's a result of something... hours and days and weeks and years of constant work and dedication."
--Roger Staubach

"You don't have to win it, just don't lose it."
--Ray Lewis to quarterback Elvis Grbac

"I just wrap my arms around the whole backfield and peel 'em one by one until I get to the ball carrier. Him I keep."
--Big Daddy Lipscomb

"Maybe a good rule in life is never become too important to do your own laundry."
--Barry Sanders

"Three things can happen when you throw the ball, and two of them are bad."
--Darrell Royal

"Football is only a game. Spiritual things are eternal. Nevertheless, Beat Texas."
--Seen on a church sign in Arkansas prior to the 1969 game

"Sure, the home-field is an advantage – but so is having a lot of talent."
--Dan Marino

"The reason women don't play football is because eleven of them would never wear the same outfit in public."
--Phyllis Diller

"People who enjoy what they are doing invariably do it well."
--Joe Gibbs

"I learned that if you want to make it bad enough, no matter how bad it is, you can make it."
--Gale Sayers

"Losing doesn't eat at me the way it used to. I just get ready for the next play, the next game, the next season."
--Troy Aikman

"Nobody who ever gave his best regretted it."
--George Halas

"I wouldn't ever set out to hurt anyone deliberately unless it was, you know, important – like a league game or something."
--Dick Butkus

"A lot of fans were drawn to me because they knew that whatever the score was, I was going to run as hard as I could on every play."
--Walter Payton

"We didn't tackle well today, but we made up for it by not blocking."
--John McKay

"I don't expect to win enough games to be put on NCAA probation. I just want to win enough to warrant an investigation."
--Bob Devaney

TELEPHONE - TEAMS - TRUST

Telephones

The telephone is an essential utility in our lives that we tend to take for granted. Telephone conversations can be powerful connections to other people or they can be distracting and confusing. How much business do you do over the phone? Are you like many of your colleagues who spend the majority of your days talking on the phone with customers. That's a tough environment to get the kind of person-to-person connection needed to capture and extend programs. With tight travel budgets, more people may be on the phone a lot more than before, so maybe some proven telephone techniques might be worth reviewing. Here are some telephone tips from Asher's "Top Ten Selling Skills":

Preparation
- Try and avoid cold calls. Someone in your company knows who you are calling or you may have a "coach" or "champion" inside their organizations. Call them first and use them as referrals.
- Do your research. Customers are pressed for time. They don't want to educate you on their organizations and programs. They want knowledgeable peers who understand their problems. Google is your ally.
- Plan your call in advance (what's the objective?), but be prepared to shift with the customer's moods and needs.

The First Call
- Introduce yourself
  o "*Good afternoon, Ms. Brown. This is [Insert Name] from [Insert Company].*"
- Grab their attention

    - *"I read in Defense News what you said about the need for better remotely piloted submersibles."*
- State the reason for the call
    - *"We're putting together a new program that we believe will revolutionize what is possible in this area."*
- Convey benefits of talking with you
    - *"We've developed a demonstrator that is easy to use and costs half as much as existing remotely piloted submersibles."*
- Request their time
    - *"Do you have a moment to discuss this?"*
- Key success factors for the call
    - Keep it short (aim for 15 seconds or less), keep it conversational (your "best friend's" voice), and keep it focused on them (why they should listen to you).
    - Open with anything but "How are you today?" and avoid first names until the customer uses yours.
    - Think before you speak, sell with emotion, paint a picture with your words.

Conversations
- Listen actively and in the moment. Multitasking is the enemy of effective listening.
- Let other people talk. Make sure your caller has completely finished speaking before responding. Sometimes they aren't done talking; they're just coming up for air.
- Practice the Q/A/F/Q technique: Ask a Question. Wait for an Answer. Feedback what was said to you to be sure you have a clear understanding of what was said. Then ask another Question to direct the conversation into the area where you want it to go. The person asking the questions controls the direction of the call.
- To create affinity with your callers, speed up or slow down your speaking voice to better match theirs. They won't realize why they feel comfortable, they just will.

- Keep in mind you can phrase anything positively, negatively or neutrally. Phrasing your words positively will help you get better results more easily.

Voice Mail
- Use the same First Call "15 Second" technique, matching the tone of their message.
- Slow your rate of speech when giving your number (try saying "area code" to give them time to write).
- Give the date and time of your call and when you can be reached or when you'll call back.
- Stay crisp. Don't keep talking until the recorder beeps and stops.
- If there's no quick response, call again the next day. After the third try, send an e-mail.

Gatekeepers
- Treat deputies and administrative assistants as decision-making customers. They usually are.
- If gatekeepers ask if they can do anything for you, let them.
- Don't make them message takers. Ask their advice on how make contact with a better time or an e-mail.
- If a customer relies on an admin, remember to copy them on e-mails. Mention their helpfulness to the customer.

Follow-up
- Craft a short e-mail to follow a telephone conversation.
  o Summarize the situation to show that you listened.
  o State what you will do and when you will do it.
  o Keep key messages in the body, not in attachments.
- Too much contact with a customer can be just as exasperating as not enough. When in doubt, be upfront and ask them how often they want contact and by what means.

The telephone can be your best friend or your worst enemy in customer communications. With a little reflection on how you've been using it, a little planning on the front end of a call, and a little adjustment in how you present your ideas and requests … you can gain a lot in building better customer relationships.

Telephone Quotes

"An amazing invention - but who would ever want to use one?"
--President Rutherford B. Hayes (making a call from Washington to Pennsylvania with the new Bell telephone)

"The bathtub was invented in 1850 and the telephone in 1875. In other words, if you had been living in 1850, you could have sat in the bathtub for 25 years without having to answer the phone."
--Bill DeWitt

"I like my new telephone, my computer works just fine, my calculator is perfect, but Lord, I miss my mind!"
--Author Unknown

"If you were going to die soon and had only one phone call you could make, who would you call and what would you say? And why are you waiting?"
--Stephen Levine

"Marriage is like a phone call in the night: first the ring, and then you wake up."
--Evelyn Hendrickson

"Utility is when you have one telephone, luxury is when you have two, opulence is when you have three – and paradise is when you have none."
--Doug Larson

"As a teenager you are at the last stage in your life when you will be happy to hear that the phone is for you."
--Fran Lebowitz

"What a lot we lost when we stopped writing letters. You can't reread a phone call."
--Liz Carpenter

"He's an honest man – you could shoot craps with him over the telephone."
--Earl Wilson

"Two of the cruelest, most primitive punishments our town deals out to those who fall from favor are the empty mailbox and the silent telephone."
--Hedda Hopper

"Someone invented the telephone, and interrupted a nation's slumbers, ringing wrong but similar numbers."
--Ogden Nash, in *Look What You Did, Christopher*

"Cell phones are the latest invention in rudeness."
--Terri Guillemets

"If it's the Psychic Network why do they need a phone number?"
--Robin Williams

"A good way to threaten somebody is to light a stick of dynamite. Then you call the guy and hold the burning fuse up to the phone. 'Hear that?' you say, 'That's dynamite, baby'."
--Jack Handy (*Saturday Night Live*)

"If The Phone Doesn't Ring, It's Me."
--song title by Jimmy Buffet

Teams

There's not a single employee in your company who doesn't depend on teams for support and success. This is especially true when you make offers to customers. Your solutions are created by teams. Your proposals are written and presented by teams.

Many lessons learned in analyses of aerospace, defense and technology losses show that that when you failed to win you likely were late-to-need on your teams. You started too late, you failed to shape the requirements, and you poorly staffed the capture team. With reduced manpower and constrained resources in this business environment, you're now even more challenged to put together the right people at the right place at the right time. Another challenge is to assign proven and full-time capture leads up front and pull in program managers and key functional leads to increase your chances for success. You're risking precious time and money if you don't gather teammates with great past performance metrics and get them working together right away.

A few words about individuals and teams. It's as true in business as in sports that there are skill players who contribute greatly to overall success. But great players cannot carry teams that are fundamentally weak. More often than not, a well-built and managed team of good players can beat opposing teams with more overall talent, but less overall cohesion.

Years ago I played a lot of rugby. To many spectators, thirty players racing up and down a rugby pitch and banging into each other looks like chaos. Not at all. Rugby is a British invention and accordingly has precise rules. Those who master the fundamentals and play as a team can beat bigger and stronger opponents. And, as I discovered, guile and experience can beat youthful enthusiasm any given Saturday.

Lessons learned in team sports, such as rugby, can help lead to success in business. For example, the former CFO of Microsoft and General Motors, Christopher Liddell, attributed his ability to tackle big ventures to his rugby experiences in his native New Zealand. This was underscored by a senior officer at J.P. Morgan Chase, the underwriter of the 2010 General Motors turnaround, who attributed Liddell's success in part to disciplined preparation he learned as a top-flight rugby player.

Baron Hanson of RedBaron Consulting, a fellow rugby enthusiast, piles on this notion by declaring that the culture of rugby parleys well into the business world. It's diverse and extremely challenging. It requires self-control and multiple skills. Rugby players are unafraid of tackling any opportunity, cannot depend on blockers, know their positions could change at any time and must earn their positions every week. Rugby players are typically both humble and loyal and, best of all, do not hold grudges.

In the early 1980s, I was one of the leaders of the San Antonio Rugby Football Club. We were diverse before it was cool. The team captain was Larry Gallegos, one of the most gifted player-leaders I've ever met. He could drop-kick a score from forty yards out on a dead run and demoralize the competition. Larry was great on the field, but not the best organizer off the field. That's where I contributed. We each knew our strengths and shortcomings and, together, made a very effective leadership team.

Besides being ethnically diverse, the San Antonio RFC was socially diverse. We had a fighter pilot and an HVAC repairman, an IBM executive and a cabinetmaker, a police officer and a parolee. We were all volunteers and we loved the game. We loved the game so much that we made a concerted effort to be competitive in the Texas Rugby Union and not be just another drinking club disguised as a sports team. One important lesson I gained from this experience that I carried into my military and business careers was that it takes a deft touch to lead a group of volunteers to success. But if you're patient and open-minded the results can exceed anyone's expectation. Teams thrive if they are treated more as volunteers than conscripts.

Because our team had a shared sense of purpose, we respected each other. There was still friction sometimes, as when a standout skill player who didn't show for the Wednesday evening practice was played at the Saturday game because we wanted to win. One of our most loyal and earnest players, rather on the small and less-gifted side, became discouraged when this happened and he didn't get to play. But when we needed a fundraiser for equipment, he ordered t-shirts made through family connections in Mexico. The shirts sold out at our annual Sevens Tournament during San Antonio Fiesta Week and we made enough money to cover our expenses for the entire year. Discouragement turned to pride and recognition (and he didn't have to buy his own beer for a year after that). Everyone on a team has value. It just may not be in expected places.

A final thought. One of the winningest rugby coaches I've ever come across, Austin Hall of Norwich University, drills into his players three simple behaviors that are worth remembering in business and life: "Get the ball, move up field, support your teammates." That Hall is the coach of the women's team should not go unmentioned.

Team Quotes

"Life is like a dogsled team. If you ain't the lead dog, the scenery never changes."
--Lewis Grizzard.

"Individuals score points, but teams win games."
--Zig Ziglar

"Teamwork is the ability to work together toward a common vision. It is the fuel that allows common people to attain uncommon results."
--Andrew Carnegie

"No man is an island, entire of itself; every man is a piece of the continent."
--John Donne

"Coming together is a beginning. Keeping together is progress. Working together is success."
--Henry Ford

"Individual commitment to a group effort – that is what makes a team work, a company work, a society work, a civilization work."
--Vince Lombardi

"The way a team plays as a whole determines its success. You may have the greatest bunch of individual stars in the world, but if they don't play together, the club won't be worth a dime."
--Babe Ruth

"No man is wise enough by himself."
--Plautus

"Michael, if you can't pass, you can't play."
--Coach Dean Smith to Michael Jordan in his freshman year at UNC

"The secret is to work less as individuals and more as a team. As a coach, I play not my eleven best, but my best eleven."
--Knute Rockne

"The team with the best players wins."
--Jack Welch

"Finding good players is easy. Getting them to play as a team is another story."
--Casey Stengel

"Remember upon the conduct of each depends the fate of all."
--Alexander the Great

"Football is an honest game. It's true to life. It's a game about sharing. Football is a team game. So is life."
--Joe Namath

"Pick good people, use small teams, give them excellent tools."
--Bill Gates

"Alone we can do so little; together we can do so much."
--Helen Keller

"Individuals play the game, but teams beat the odds."
--SEAL Team saying

"When he took time to help the man up the mountain, lo, he scaled it himself."
--Tibetan Proverb

"Which team to use for what purpose is a crucial, difficult and risky decision that is even harder to unmake. Managements have yet to learn how to make it."
--Peter F. Drucker

"A single arrow is easily broken, but not ten in a bundle."
--Japanese proverb

"What we need to do is learn to work in the system, by which I mean that everybody, every team, every platform, every division, every component is there not for individual competitive profit or recognition, but for contribution to the system as a whole on a win-win basis."
--W. Edward Deming

"A team is where a boy can prove his courage on his own. A gang is where a coward goes to hide."
--Mickey Mantle

"Light is the task where many share the toil."
--Homer

"No member of a crew is praised for the rugged individuality of his rowing."
--Ralph Waldo Emerson

"A major reason capable people fail to advance is that they don't work well with their colleagues."
--Lee Iacocca

"If I could solve all the problems myself, I would."
--Thomas Edison, when asked why he had a team of twenty-one assistants

"We must indeed all hang together, or, most assuredly, we shall all hang separately."
--Benjamin Franklin

"In union there is strength."
--Aesop

"It is better to have one person working with you than three people working for you."
--Dwight D. Eisenhower

"A boat doesn't go forward if each one is rowing their own way."
--Swahili proverb

"You can't be a real country unless you have a beer and an airline. It helps if you have some kind of a football team, or some nuclear weapons, but at the very least you need a beer."
--Frank Zappa

"Fit no stereotypes. Don't chase the latest management fads. The situation dictates which approach best accomplishes the team's mission."
--Colin Powell

"You can't just beat a team, you have to leave a lasting impression in their minds so they never want to see you again."
--Mia Hamm

Trust

It seems like trust is on every senior leader's briefing slide listing the things needing improvement. While customer trust is always vital, it's especially important during times of flat or declining markets. Will your current customers trust you enough to stay with you in the coming years? Will new customers trust you more than your competitors?

Considering this industry sector I think it's important at the outset to recognize that, like it or not, you cannot "engineer" customer trust. Trust is not a material thing; it's a human feeling. But that's not to say it's intangible and you can't do anything about it. You can earn it by your actions. In *The Speed of Trust,* author Stephen M. R. Covey equates trust with confidence (and distrust with suspicion) and promotes the concept that if you can engender trust there are real economic benefits to be gained. Covey's observation is that trust always affects two business outcomes – speed and cost. When trust goes down, speed goes down and cost goes up. But when trust goes up, speed goes up and cost goes down.

Another favorite book on the subject is *The Trusted Advisor*, by David Maister. His definition of a trusted advisor is the person a client turns to when an issue first arises and with whom virtually everything is open for discussion. While there are a lot of useful insights in this book about developing better business relationships, I especially like the list of four essential elements that engender trust: credibility, reliability, intimacy, and a low level of self-orientation.

When customers trust a company it's the people they've dealt with who have made the difference, rather than the product or solution that's being delivered. Ultimately, business is deeply personal. Every human transaction counts. Why would any customer pick your company? It's hard to trust a corporation. But a customer can trust you.

Trust Quotes

"I think that we may safely trust a good deal more than we do."
--Henry David Thoreau

"Our distrust is very expensive."
--Ralph Waldo Emerson

"One must be fond of people and trust them if one is not to make a mess of life."
--E.M. Forster

"We're all born brave, trusting, and greedy, and most of us remain greedy."
--Mignon McLaughlin

"Few things help an individual more than to place responsibility upon him, and to let him know that you trust him."
--Booker T. Washington

"In God we trust, all others we virus scan."
--Author Unknown

"Every two years the American politics industry fills the airwaves with the most virulent, scurrilous, wall-to-wall character assassination of nearly every political practitioner in the country – and then declares itself puzzled that America has lost trust in its politicians."
--Charles Krauthammer

"Trust your own instinct. Your mistakes might as well be your own, instead of someone else's."
--Billy Wilder

"As soon as you trust yourself, you will know how to live."
--Johann Wolfgang Von Goethe

"Love all, trust a few, do wrong to none."
--William Shakespeare

"Deciding whether or not to trust a person is like deciding whether or not to climb a tree, because you might get a wonderful view from the highest branch, or you might simply get covered in sap, and for this reason many people choose to spend their time alone and indoors, where it is harder to get a splinter."
--Lemony Snicket

"A wedding anniversary is the celebration of love, trust, partnership, tolerance and tenacity. The order varies for any given year."
--Paul Sweeney

"Do not trust all men, but trust men of worth; the former course is silly, the latter a mark of prudence."
--Democritus

"You may be deceived if you trust too much, but you will live in torment if you do not trust enough."
--Frank Crane

"It is an equal failing to trust everybody, and to trust nobody."
--English proverb

"Never trust the advice of a man in difficulties."
--Aesop

"The people I distrust most are those who want to improve our lives but have only one course of action."
--Frank Herbert

"The only way to make a man trustworthy is to trust him."
--Henry L. Stimson

"Never trust anything that can think for itself if you can't see where it keeps its brain."
--J. K. Rowling

"Nobody believes the official spokesman but everybody trusts an unidentified source."
--Ron Nessen

"We are inclined to believe those whom we do not know because they have never deceived us."
--Samuel Johnson

"The older I grow the more I distrust the familiar doctrine that age brings wisdom."
--H.L. Mencken

"I learned you have to trust yourself, be what you are, and do what you ought to do the way you should do it."
--Barbra Streisand

"Trust one who has gone through it."
--Virgil

"Never trust the teller. Trust the tale."
--D.H. Lawrence

"Never trust the man who tells you all his troubles but keeps from you all his joys."
--Jewish proverb

"Mistrust the man who finds everything good, the man who finds everything evil, and still more the man who is indifferent to everything."
--Johann K. Lavater

"Put more trust in nobility of character than in an oath."
--Solon

"Trust everybody, but cut the cards."
--Finley Peter Dunne

"Don't trust anyone over 30."
--Jerry Rubin

"It is impossible to go through life without trust: that is to be imprisoned in the worst cell of all, oneself."
--Graham Greene

"Put not your trust in money, but put your money in trust."
--Oliver Wendell Holmes

U

UNDERSTANDING BUYER-SELLER ALIGNMENT

Understanding Buyer-Seller Alignment

Greater understanding of what goes on between buyers and sellers at various stages can be helpful in gaining new orders. In section C there's a piece about the need for challenging some customers. Research by Mathew Dixon and Brent Adamson of the Sales Executive Council, a program within the Corporate Executive Board, showed how buying patterns of large-account customers have changed over the years. Instead of asking for products, these administrative customers began seeking solutions and their buying behaviors became more complex. They demanded broad-based consensus among stakeholders, increasing risk aversion on the part of key decision makers, and calling for greater customization. Their buying cycle shifted.

You can have a solid sales process (section P) and talented salespeople and still lose business. Why? A major reason could be that your sales process isn't aligned to your customer's buying cycle. Just as there's no one-size-fits-all sales process for companies, there's not a standard buying cycle for customers. But, as you can observe from your own buying behavior, there is a beginning and an end with things happening in between.

Here is a notional buying cycle:

Awareness > Interest > Research > Desire > Risk Alleviation > Purchase

And here is a notional sales process:

Prospect > Interview > Analyze Needs > Present > Negotiate > Close

If your sales process is in sync with your customer's buying cycle, it makes a sale much more likely. But for a variety of reasons, such as greater availability of technical information and urgency on the part of government acquisition officials to squeeze more into declining budgets, your customers are often way out in front of you (in the Desire phase) when you make your first calls (in the Interview phase). And it irritates them. You're there to uncover their "pain" and start thinking of solutions and they've already had enough pain and are making acquisition plans.

Just selling "solutions" may not be enough for many of your customers. The same authors of *The Challenger Sale* (See Challenge Your Customer) published an interesting paper in the Harvard *Business Review* entitled "The End of Solution Sales." They observe that typical Solution Selling, where polished sales professionals discover customer needs and build solutions, are good for customers who have a good understanding of what their problems are, but don't know how to proceed. With more sophisticated procurement teams backed by purchasing consultants, a lot of customers now believe they can readily define solutions for themselves. The authors found that, on average, many large account customers felt they had completed 60 percent of their purchasing decision before even having a discussion with a supplier.

So even if you can get ahead of the RFP, it still may be a struggle to close the customer and avoid a price shoot-out. What can you do? Perhaps you could change the game with some behaviors from *The Challenger Sale*. Rather than using sales calls to uncover pain or ferreting out RFP details, reframe the discussion and turn the already-decided customer into one with emerging needs (i.e. slow down the buying cycle to help them get a better solution). You could branch out and focus more on agile customer organizations that are in a state of flux (needing a faster, more creative transaction) and provide them with innovative, lower-cost solutions. You could seek out and win over different stakeholders (skeptical change agents rather than the usual friendly informants) and coach your customers and stakeholders on how to buy smarter. This is what the authors call Insight Selling.

At the very least you need to constantly ask yourself, "Are my customers buying what I'm selling?" If you have doubts, maybe it's time to take your sales process into the shop for alignment. While you're waiting for it to get fixed you'll have time to consider all that goes into understanding customers. Here are some quotes to get you kick-started.

Understanding the Customer Quotes

"The beginning is the most important part of the work."
--Plato

"The best vision is insight."
--Malcolm S. Forbes

"You can observe a lot just by watching."
--Yogi Berra

"The power of accurate observation is commonly called cynicism by those who have not got it."
--George Bernard Shaw

"Sometimes the questions are complicated and the answers are simple."
--Theodore "Dr. Seuss" Geisel

"I think the one lesson I have learned is that there is no substitute for paying attention."
--Diane Sawyer

"There is only one boss. The customer. And he can fire everybody in the company, from the chairman on down, simply by spending his money somewhere else."
--Sam Walton

"Your most unhappy customers are your greatest source of learning."
--Bill Gates

"If you have always done it that way, it is probably wrong."
--Charles Kettering

"There is nothing so useless as doing efficiently that which should not be done at all."
--Peter F. Drucker

"Common sense ain't common."
--Will Rogers

"Everyone is entitled to their own opinion, but not their own facts."
--Daniel Patrick Moynihan

"The cure for boredom is curiosity. There is no cure for curiosity."
--Dorothy Parker

"Education is a progressive discovery of our own ignorance."
--Will Durant

"Learning without thought is labor lost."
--Confucius

"An organization's ability to learn, and translate that learning into action rapidly, is the ultimate competitive advantage."
--Jack Welch

"Predicting is difficult – especially about the future."
--Victor Borge

"One of the best rules in conversation is never to say a thing which any of the company can reasonably wish had been left unsaid."
--Jonathan Swift

"Great minds discuss ideas, average minds discuss events, small minds discuss people."
--Admiral Hyman Rickover

"There are two types of people in this world: good and bad. The good sleep better, but the bad seem to enjoy the waking hours much more."
--Woody Allen

"Trust yourself. You know more than you think you do."
--Benjamin Spock

"Nothing at all will be attempted if all possible objections must first be overcome."
--Samuel Johnson

"You may not win them over, but if you hang around long enough, you'll wear them out."
--Phil Simms

"Discovery consists in seeing what everyone else has seen but understanding it for the first time."
--Albert Szent-Gyorgyi

"The difficulty lies, not in the new ideas, but in escaping the old ones, which ramify, for those brought up as most of us have been, into every corner of our minds."
--John Maynard Keynes

"You cannot depend on your eyes when your imagination is out of focus."
--Mark Twain

"A problem is a chance for you to do your best."
--Duke Ellington

"Fortune befriends the bold."
--John Dryden

"Charm is a way of getting the answer yes without asking a clear question."
--Albert Camus

"As long as you're going to be thinking anyway, think big."
--Donald Trump

"Patience is the companion of wisdom."
--St. Augustine

"Learning from experience is a faculty almost never practiced."
--Barbara Tuchman

"Anything that won't sell, I don't want to invent. Its sale is proof of utility, and utility is success."
--Thomas Edison

"There's no praise to beat the sort you can put in your pocket."
--Moliere

"You lose, you smile, and you come back the next day. You win, you smile, you come back the next day."
--Ken Griffey Jr.

VOICE OF THE CUSTOMER - VOICE TONE - VIDEOS

Voice of the Customer

When you hear "Voice of the Customer," what do you think of? In the commercial business world this is a market research technique to identify a detailed set of customer wants and needs that are then prioritized in terms of relative importance and satisfaction with current alternatives. This information is gathered through interviews, discussions, inquiries, etc. Successful companies involve product development core teams early on in this process, making it a valuable starting point for innovation.

Perhaps you've also heard the term "Customer Experience Management." Too often this is used as an expression for activities attempting to influence customers to do what you want them to do. But in its classic form it's more complex. Customer Experience Management has two components: Voice of the Customer and Customer Value Analysis. Voice of the Customer refers to understanding customer needs and desires in order to provide products and services to satisfy those needs and desires. Customer Value Analysis refers to the methods and tools, such as Value Proposition modeling, needed to ensure that those needs and desires are satisfied in ways that provide your company with competitive advantage. It's ultimately all about value in the eyes of the customer, the measure of "worth what's paid for" – a trade-off usually between quality and price. Value propositions are essentially promises you make to customers (see Promises).

The needs obtained through Voice of the Customer are of two types: stated (spoken) and latent (unspoken). Stated needs are those that customers are willing and able to articulate in an interview or group discussion. Latent needs are those that customers do not state, either because they think that they are obvious, or because they themselves don't know about these needs. Latent needs are normally identified through observational analysis rather than through inquiries. Customer needs data can be qualitative or quantitative – both are measurable. For example, a key decision maker might not ever want to have to testify again to congress on a failed program or might want a program to come in on time and on budget to meet a key warfighter need.

Have you ever seen Voice of the Customer inputs related to your business sector? Every year at TechAmerica's Vision Conference, volunteer groups from defense, aerospace, and technology companies provided independent forecasts of their markets. In their presentations they also included non-attributed "messages to industry" gained through interviews with government acquisition officials. These can be insightful. Here are some Voice of the Customer quotes I've come across over the past five years from these interviews. Not surprisingly, they do not change much over time. Are you listening to what they're saying?

Voice of the Customer Quotes

Don'ts

"Don't bother me with marketing your irrelevant offerings."

"Don't bring a point solution, but a range of solutions."

"Don't market something that has had no processes or rigor applied."

"Don't just sell cost, schedule, performance."

"Don't oversell capabilities, delivery times, and low costs."

"Don't bring us things to add to our inventory. Bring things that replace current inventory."

"Don't expect large, complex development contracts – focus on commercial off-the-shelf (COTS) and adding/integrating security."

"Don't be a stranger. We want to hear about innovations."

Irritants

"Do your homework! Get smart on our programs before coming to us. Hire knowledgeable subject matter experts to make you smart."

"Understand our mission and know how it is evolving."

"Industry needs to approach us as one enterprise and not as multiple, individual customers."

"Industry should bring solutions for us to evaluate against mission needs, but industry needs to know our needs first."

"The proposals all read alike – little to no differentiation discernible. We don't have the expertise to read between the lines."

"Tell us your story, not a litany of mind numbing facts. Show us your value."

"Data must be king, but we're overwhelmed with the volume of data. We can't efficiently get executable information out of it."

"Provide more certified or certifiable products, not more proprietary systems."

"Industry must deliver what it signs up to."

"Major programs tend to be substantially late, over budget, and don't meet requirements."

"Industry needs to plan and execute for the long term and spend less on short-term gains."

"You are only as good as your last flight test mission."

Desires

"Help us understand the problems and help us identify the left and right limits of the solution set."

"Help us achieve our missions within the budgets we will have."

"We want to partner with industry, but at the lowest price possible with industry taking on more risk and anticipating what we really want."

"We cannot keep up with technology. How can we better push information where needed? How do we become more like an iPhone?"

"We need to strive for commonality across our programs, wherever possible."

"We underestimate the difficulty and cost of integrating systems; industry needs to help educate us."

"We need to collectively improve the outcomes more than the processes."

"Government and industry need a win-win where affordable solutions are profitable."

"We need more dialog with industry to tell us what is driving their costs up."

"Anticipate changes to the programs of record. Be ready to offer alternative solutions."

"Provide a solid business case for adding or evolving capabilities."

"We want more cross-industry solutions, consortia, and idea sharing."

"Continue to enhance capability while making the weapons system more user friendly and cost effective."

"Perform, perform, perform! Current and past performance will help us reduce the risks for contracted efforts."

"Sometimes the program hasn't failed; you just failed to manage expectations."

Voice Tone

According to research, voice tone can convey up to 38 percent of the content of oral communication (See X-ray Vision). When you think about it, this isn't surprising. You can change the meaning of most sentences just by altering the inflection in your voice and shifting emphasis from one word to another. Your voice gives emotion to what you say. But what is your emotion (or lack thereof) telling your customers? Were you ever in a conversation with someone and their tone of voice was saying something different than the words they were saying? Was it arrogance you felt? Indifference? Sarcasm? It didn't feel good, did it?

Now think of a conversation you've experienced in which the words were ordinary and expected, but the speaker's tone of voice gave you calm assurance and a feeling of trust. It could have been with any number of people, but it likely was with a close colleague or friend, right? This is a technique taught during Corporate Visions' Power Messaging course. If you want to positively influence your customer, use your "best friend's voice." Who wouldn't trust their best friend? Who wouldn't buy from their best friend?

Another great voice tone technique comes from customer relations guru Dr. Tom Barrett. He suggests calling your own voice mail to hear what your tone is saying to those who want to get in touch with you. Do you sound disinterested and unapproachable? Or do you sound upbeat and eager to hear from them? It only takes a few seconds to change the recording and improve your impression in the ears of your callers.

Words are important. Body language is important. But don't forget the impact your tone of voice can have. Mom was right, it's often not what you say that matters. It's how you say it.

Voice Tone Quotes

"We often refuse to accept an idea merely because the tone of voice in which it has been expressed is unsympathetic to us."
--Friedrich Nietzsche

"Don't look at me in that tone of voice."
--Gene Watson

"The quietness of his tone italicized the malice of his reply."
--Truman Capote

"All feelings have their peculiar tone of voice, gestures and looks, and this harmony, as it is good or bad, pleasant or unpleasant, makes people agreeable or disagreeable."
--François de la Rochefoucauld

"We are not won by arguments that we can analyze, but by the tone and temper, by the manner which is the man himself."
--Louis Brandeis

"Take the tone of the company that you are in."
--Lord Chesterfield

"Words mean more than what is set down on paper. It takes the human voice to infuse them with deeper meaning."
--Maya Angelou

"To achieve the plain, even tone recommended by Whitman and Mark Twain, the first requisite is sincerity; the second is a distinct thought."
--Jacques Barzun

" 'When I use a word,' Humpty Dumpty said in rather a scornful tone, 'it means just what I choose it to mean – neither more nor less'."
--Lewis Carroll

"Speak when you are angry – and you'll make the best speech you'll ever regret."
--Dr. Laurence J. Peter

"Kind words can be short and easy to speak, but their echoes are truly endless."
--Mother Teresa

"Good communication does not mean that you have to speak in perfectly formed sentences and paragraphs. It isn't about slickness. Simple and clear go a long way."
--John Kotter

"The tongue is the only tool that gets sharper with use."
--Washington Irving

"Speech is the mirror of the soul; as a man speaks, so is he."
--Publilius Syrus

"I feel that if a person has problems communicating, the least he can do is shut up."
--Tom Lehrer

Videos

You're in a long meeting and all the words in the presentation are blending together in white noise. You stealthily look at your watch. Time seems to have stopped. You think of being anywhere but in the room. All of a sudden the speaker introduces a short video and you pop back into focus.

What you experienced was a shift in media stimulus that helped your brain refresh and reconnect. For many years educators have known that instructional videos help students understand concepts more rapidly, retain more information, and become more enthusiastic in the process. Videos have the power to deliver lasting images and can convert your presentations from boring to memorable.

During the customer relations classes I teach, I've often noted the huge impact videos have on the mood of the room. I've found that well-timed (and humorous) videos wake up the audience and reinforce key messages. When delivering the same presentations in rooms without audiovisual support or over live meetings and teleconferences, it's just not the same.

I believe there are four main things to consider in successfully using videos in presentations:

Selection

For a video to be effective it has to be appropriate for both the message and the audience, and it should illustrate the point you want to make. Too short and it's a distraction. Too long and you lose the impact. In training classes I often employ "wakeup" videos. There's always a way to connect them to teaching points with a few well-chosen and playful remarks. A word of caution: showing a video just to show a video may be entertaining on one level, but it's really a wasted opportunity to reinforce your message.

Placement

When is the best time to show the video? Before, during, or after your presentation? It depends, of course, on many variables. The optimal place is where it will have the best "wow" factor. For example, showing what a product will actually "do" after you've displayed some performance figures or in closing the presentation for emotional connection.

Preview

Always test the video before you begin the presentation. Every meeting room has its own audiovisual dysfunctionality and it's better to find out beforehand that the sound system isn't on or the DVD player isn't hooked up. Putting on the video with "I hope this works" doesn't engender confidence from your audience.

Introduction

Be sure to make a short, meaningful introduction to your video so that you don't surprise and confuse your audience. Remember, some of them may be in sleep mode. Make a strong link to what you were talking about or make a smooth transition to what you will be talking about. You went through all the trouble to bring the video. Make it count.

Since we tend to live out our lives in movies, here are some favorite movie quotes you can use to liven things up.

Movie Quotes

"You don't understand! I coulda had class. I coulda been a contender. I could've been somebody, instead of a bum, which is what I am."
--*On the Waterfront*

"Go ahead, make my day."
--*Sudden Impact*

"All right, Mr. DeMille, I'm ready for my close-up."
--*Sunset Boulevard*

"Fasten your seatbelts. It's going to be a bumpy night."
--*All About Eve*

"What we've got here is failure to communicate."
--*Cool Hand Luke*

"There's no place like home."
--*The Wizard of Oz*

"Show me the money!"
--*Jerry Maguire*

"You can't handle the truth!"
--*A Few Good Men*

"I want to be alone."
--*Grand Hotel*

"After all, tomorrow is another day!"
--*Gone with the Wind*

"Round up the usual suspects."
--*Casablanca*

"I'll have what she's having."
--*When Harry Met Sally*

"You're gonna need a bigger boat."
--*Jaws*

"Badges? We ain't got no badges! We don't need no badges! I don't have to show you any stinking badges!"
--*The Treasure of the Sierra Madre*

"Today, I consider myself the luckiest man on the face of the earth."
--*The Pride of the Yankees*

"If you build it, he will come."
--*Field of Dreams*

"We rob banks."
--*Bonnie and Clyde*

"Plastics."
--*The Graduate*

"I see dead people."
--*The Sixth Sense*

"Well, nobody's perfect."
--*Some Like it Hot*

"It's alive! It's alive!"
--*Frankenstein*

"Houston, we have a problem."
--*Apollo 13*

"You shut your mouth when you're talking to me!"
--*The Wedding Crashers*

"Roses are red, violets are blue, I'm a schizophrenic and so am I."
--*What about Bob?*

"You might have seen a housefly, maybe even a superfly, but I bet you ain't never seen a donkey fly!"
--*Shrek*

"I'm not bad, I'm just drawn that way."
--*Who Framed Roger Rabbit*

Reporter: "Tell me, how did you find America?" John Lennon: "Turned left at Greenland."
--*A Hard Day's Night*

"There's no reason to become alarmed, and we hope you'll enjoy the rest of your flight. By the way, is there anyone on board who knows how to fly a plane?"
--*Airplane*

"Name's Barf. I'm a Mog, half man half dog. I'm my own best friend."
--*Spaceballs*

"The key here, I think, is to not think of death as an end. But, but, think of it more as a very effective way of cutting down on your expenses."
--*Love and Death*

"It's amazing the clarity that comes with psychotic jealousy."
--*My Best Friend's Wedding*

"If I'm not back in five minutes... wait longer!"
--*Ace Ventura: Pet Detective*

"It's 106 miles to Chicago, we've got a full tank of gas, half a pack of cigarettes, it's dark, and we're wearing sunglasses."
--*The Blues Brothers*

"Hope is a good thing, maybe the best of things, and no good thing ever dies."
--*The Shawshank Redemption*

"Ray, next time someone asks you if you're a god, you say YES!"
--*Ghostbusters*

"SQUIRREL!"
--*Up*

"That's just the way it crumbles – cookie-wise."
--*The Apartment*

"Gentlemen, you can't fight in here! This is the War Room!"
--*Dr. Strangelove*

"Are you crazy? The fall will probably kill you."
--*Butch Cassidy and the Sundance Kid*

"I'm hungry. Let's get a taco."
--*Reservoir Dogs*

"They've done studies, you know. Sixty percent of the time, it works every time."
--*Anchorman*

"Demented and sad, but social."
--*The Breakfast Club*

"I don't understand. All my life I've been waiting for someone and when I find her, she's… she's a fish."
--*Splash*

"Go that way, really fast. If something gets in your way, turn."
--*Better Off Dead*

"Sometimes you win, sometimes you lose, and sometimes it rains."
--*Bull Durham*

"Hokey religions and ancient weapons are no match for a good blaster by your side, kid."
--*Star Wars*

"I'm going to make him an offer he can't refuse."
--*The Godfather*

"My life is as good as an Abba song. It's as good as Dancing Queen."
--*Muriel's Wedding*

"This is the guy behind the guy behind the guy."
--*Swingers*

"All I've ever wanted was an honest week's pay for an honest day's work."
--*Bilko*

"Wait Master, it might be dangerous... you go first."
--*Young Frankenstein*

"How did the pig tracks get on the ceiling?"
--*The Simpsons Movie*

"Read between the lines, Theo. Read between the lines!"
--*School of Rock*

"It was a good plan, up till now."
--*Pirates of the Caribbean*

"I'll be back."
--*The Terminator*

"My Mama always said, 'Life was like a box of chocolates; you never know what you're gonna get'."
--*Forrest Gump*

"You talkin' to me? You talkin' to me? You talkin' to me?"
--*Taxi Driver*

"Fuh-get about it!"
--*Donnie Brasco*

"Dave, this conversation can serve no purpose anymore. Goodbye."
--*2001: A Space Odyssey*

WORDS - WISE WORDS

Words

According to research, words make up only seven percent of oral communication (See Voice Tone and X-Ray Vision). But what an important seven percent. Have you ever been in a meeting or situation where one perfect word said at a perfect moment changed everything? It's stunning how something so small can have such a large impact.

Words are important. We want to use the right words with the right customers at the right time. Key decision makers are more interested in words about value and meaning, while technical buyers seek more precise, data-driven words. Operators want words about performance and funders respond best to words linked to costs and schedules. While our value propositions are the same for each customer set, we continually need to remind ourselves that the value messages for each type of customer need to be adjusted with words reflecting each customer's world, each customer's needs.

Always on the lookout for helpful lists of things to consider when communicating with customers, I came across the *Ten Rules of Successful Communication* by Frank Luntz. He's a political consultant with a *New York Times* Best Seller: *Words That Work: It's Not What You Say, It's What People Hear*. Luntz claims the tactical use of words affects what we buy, who we vote for and what we believe. Here's an adaptation of his list of important considerations in selecting your words:

1. Simplicity is best
2. Brevity has value
3. Credibility matters
4. Consistency makes it memorable
5. Novelty captures the imagination
6. Sound and Texture reinforce meaning

7. Humanized messages convey emotionally
8. Visual images last longer
9. Questions set up answers (the rhetorical question is as old as Socrates and as fresh as "Got milk?")
10. Relevant context gives meaning

And here's some more expert advice on choosing your words:

Word Quotes

"For just when ideas fail, a word comes in to save the situation."
--Johann Wolfgang von Goethe

"Broadly speaking, the short words are the best, and the old words are the best of all."
--Winston Churchill

"Be careful of the words you say, keep them short and sweet. You never know, from day to day, which ones you'll have to eat."
--Anonymous

"Don't use words too big for the subject. Don't say 'infinitely' when you mean 'very'; otherwise you'll have no word left when you want to talk about something really infinite."
--C. S. Lewis

"The difference between the right word and the almost right word is the difference between lightning and a lightning bug."
--Mark Twain

"As we must account for every idle word, so must we account for every idle silence."
--Benjamin Franklin

"But words are things, and a small drop of ink, falling like dew, upon a thought, produces that which makes thousands, perhaps millions, think."
--George Gordon Byron

"Boy, those French: they have a different word for everything!"
--Steve Martin

"Colors fade, temples crumble, empires fall, but wise words endure."
--Edward Thorndike

"A great many people think that polysyllables are a sign of intelligence."
--Barbara Walters

"I like good strong words that mean something."
--Louisa May Alcott

"What's another word for Thesaurus?"
--Steven Wright

"If you talk to a man in a language he understands, that goes to his head. If you talk to him in his language, that goes to his heart."
--Nelson Mandela

"Words – so innocent and powerless as they are, as standing in a dictionary, how potent for good and evil they become, in the hands of one who knows how to combine them!"
--Nathaniel Hawthorne

"The six most important words: I admit I made a mistake. The five most important words: You did a good job. The four most important words: What is your opinion? The three most important words: If you please. The two most important words: Thank you. The one least important word: I."
--Anonymous

"So difficult it is to show the various meanings and imperfections of words when we have nothing else but words to do it with."
--John Locke

"Words are not as satisfactory as we should like them to be, but, like our neighbors, we have got to live with them and must make the best and not the worst of them."
--Samuel Butler

"Truthiness is tearing apart our country."
--Stephen Colbert

"A word is dead when it is said, some say. I say it just begins to live that day."
--Emily Dickinson

"Suit the action to the word, the word to the action."
--William Shakespeare

"If we can boondoggle ourselves out of this depression, that word is going to be enshrined in the hearts of the American people for years to come."
--Franklin D. Roosevelt

"Civilization began the first time an angry person cast a word instead of a rock."
--Sigmund Freud

"Bill Gates is a very rich man today... and do you want to know why? The answer is one word: versions."
--Dave Barry

"You can get much farther with a kind word and a gun than you can with a kind word alone."
--Al Capone

"A word once uttered can never be recalled."
--Horace

"Every once in a while, you let a word or phrase out and you want to catch it and bring it back. You can't do that. It's gone, gone forever."
--Dan Quayle

"Speak clearly, if you speak at all; carve every word before you let it fall."
--Oliver Wendell Holmes

"Always give a word or sign of salute when meeting or passing a friend, or even a stranger, if in a lonely place."
--Tecumseh

Wise Words

In the introduction to this book I mentioned my father the paint salesman. Only later in life did I realize his business success didn't come from just selling paint, but forging customer relationships. George Potts passed away at the age of 90. As I was going through Dad's papers seeking inspiration for what I needed to say at his memorial service, I came across a letter he sent to my mother during World War II. He wrote that during training at Fort Sheridan he'd learned four important rules of military life he felt might be of use even if he survived the war:

- Salute everyone and let them sort it out.
- Never break into a chow line.
- Don't volunteer that you have special talents.
- Avoid causing a disturbance you could be blamed for.

Wise words indeed. But for the most part these will only help you stay out of trouble. What about when it's time to move forward and make a sale?

Before George died he gave a lot of advice to anyone willing to listen. Once my nephew Mike was visiting George in the nursing home. Mike's a medical equipment salesman in California. He asked his grandfather, "Gramps, do you have any advice on how I can get ahead of the competition?" Without hesitation George replied, "Give your customers a firm handshake, look them straight in the eyes, and smile. At that moment you're already ahead of your competitors." Mike reported that since he concentrated on these three simple things his sales were up despite a soft market. Maybe you too might find them useful.

Wisdom Quotes

"The wisdom of the wise and the experience of the ages are perpetuated by quotations."
--Benjamin Disraeli

"By three methods we may learn wisdom: first, by reflection, which is noblest; second, by imitation, which is easiest; and third by experience, which is the bitterest."
--Confucius

"Wisdom is the quality that keeps you from getting into situations where you need it."
--Doug Larson

"The only true wisdom is in knowing you know nothing."
--Socrates

"A wise man is superior to any insults which can be put upon him, and the best reply to unseemly behavior is patience and moderation."
--Moliere

"Knowledge is proud that it knows so much; wisdom is humble that it knows no more."
--William Cowper

"Honesty is the first chapter in the book of wisdom."
--Thomas Jefferson

"A man begins cutting his wisdom teeth the first time he bites off more than he can chew."
--Herb Caen

"Every man is a damn fool for at least five minutes every day; wisdom consists in not exceeding the limit."
--Elbert Hubbard

"Be happy. It's one way of being wise."
--Sidonie Gabrielle Colette

"Wisdom is the supreme part of happiness."
--Sophocles

"The art of being wise is the art of knowing what to overlook."
--William James

"From the errors of others, a wise man corrects his own."
--Publilius Syrus

"Wisdom is the reward you get for a lifetime of listening when you'd have preferred to talk."
--Doug Larson

"Turn your wounds into wisdom."
--Oprah Winfrey

"There is a wisdom of the head, and a wisdom of the heart."
--Charles Dickens

"Cleverness is not wisdom."
--Euripides

"Ignorant men raise questions that wise men answered a thousand years ago."
--Johann Wolfgang von Goethe

"Common sense in an uncommon degree is what the world calls wisdom."
--Samuel Taylor Coleridge

"Nobody can give you wiser advice than yourself."
--Marcus Tullius Cicero

"Patience is the companion of wisdom."
--Saint Augustine

"It requires wisdom to understand wisdom: the music is nothing if the audience is deaf."
--Walter Lippmann

"The doorstep to the temple of wisdom is a knowledge of our own ignorance."
--Benjamin Franklin

"It is a characteristic of wisdom not to do desperate things."
--Henry David Thoreau

"Wise sayings often fall on barren ground, but a kind word is never thrown away."
--Arthur Helps

"Better be wise by the misfortunes of others than by your own."
--Aesop

"To conquer fear is the beginning of wisdom."
--Bertrand Russell

"A prudent question is one-half of wisdom."
--Francis Bacon

"Knowledge can be communicated, but not wisdom. One can find it, live it, be fortified by it, do wonders through it, but one cannot communicate and teach it."
--Hermann Hesse

"Some folks are wise and some are otherwise."
--Tobias Smollett

"Wisdom doesn't necessarily come with age. Sometimes age just shows up all by itself."
--Tom Wilson

"In seeking wisdom thou art wise; in imagining that thou hast attained it – thou art a fool."
--Lord Chesterfield

"Wisdom is not wisdom when it is derived from books alone."
--Horace

"Memory is the mother of all wisdom."
--Aeschylus

"No man was ever wise by chance."
--Lucius Annaeus Seneca

"Nine-tenths of wisdom is being wise in time."
--Theodore Roosevelt

"Knowledge comes, but wisdom lingers."
--Alfred Lord Tennyson

X-RAY VISION

X-Ray Vision

Did you ever wish you had Superman's x-ray vision? *Stories in Action* Comics in the late 1930's showed Superman with the ability to see through objects. It didn't really work like an x-ray, it was more like psychic remote viewing – the ability to see what others can't and uncover something of great importance. In that sense, we all have x-ray vision, but we don't flip it on very often. Let's focus on customers. It's difficult to see into their minds, but we can observe their physical movements. Body language can reveal more of the customer's emotional state and intentions than words and voice tone can. However, most of the time we aren't looking while we're listening.

In landmark research at UCLA in the 1960s, Professor Albert Mehrabian provided evidence that feelings and attitude are communicated seven percent in spoken words, 38 percent in how words are said and 55 percent in body expression. When we actively listen to a customer we tend to concentrate on the 45 percent of the oral communication we can hear: words and tone. It's a mistake to ignore the language of the body. I remember when the classic paperback *Body Language* by Julius Fast came out in 1970. Every guy I knew bought it because it purportedly taught you how to be more successful with women. And likely we were, if for no other reason than after reading the book we started being more attentive to the ladies. That's the secret to your x-ray vision: you don't have to be a body language expert to read people. You just have to pay attention. It's so simple even your pets can do it.

It's been documented that the body expresses the mind. We gesture before we learn to talk and gesture even if we are blind or on the phone, suggesting that body language is not necessarily intended for the recipient. Although the study of body language (kinesics) is still said to be in development, it's not something new. Hippocrates, Cicero, Francis Bacon, and Charles Darwin wrote of nonverbal communication, but it's always been seen as a sideshow to the importance of words and tone. Perhaps this is because body language is difficult to define and measure. It's both genetically and environmentally determined and has vast variations by personality, gender, culture, etc. Yet we can easily read happiness, sadness, fear, disgust, surprise and anger in faces.

So where should we direct our x-ray vision? The body language experts recommend searching three major areas: face and eyes, arms and hands, legs and posture. The key is to look for tension. Generally speaking, tension in a body part is a red flag. For example, is the brow smooth or furrowed and are the eyes in contact or in avoidance? Are the arms relaxed or crossed and are the hands open or clenched? Are the legs still or bouncing and is the body moving toward or away from you? If you want more specifics and opinions, there are a great number of books and blogs on the subject. If you're doing business internationally, remember there are cultural differences: the body language of the Japanese compared to the Italians, for example. All said, if you look at multiple body gestures and put them in context with words, voice tone and settings, you'll get a much better picture of what the customer is thinking and feeling.

Should you only have a brief encounter with a customer or can only muster enough concentration to focus on one body part, you probably want to focus on the eyes. The eyes say so much. We can sense eye contact at great distances and often have a feeling when someone is watching us. We can detect whether eyes recognize us as another human being or merely an object, whether they are cold or warm, confident or furtive. From analysis called neuro-linguistic programming (NLP) developed in the 1960s, it's said that during speech the eyes look right when the brain is imagining and left when it is recalling. Within those two spatial areas, eyes up indicate visual learners, eyes level indicate auditory learners and eyes downward indicate kinesthetic ("doing") learners. These can be considerations for how you may want to deliver information to your customer. Further, widened eyes show interest, while narrowed eyes show concern. An "eyebrow flash" (quick raise and lower) is a friendly sign of greeting and acknowledgement, while fully raised eyebrows can show surprise, disbelief and skepticism. Rapid blinking shows excitement or pressure (and sometimes lying). And don't forget that in some cultures, avoiding eye contact shows respect, while direct eye contact is, pardon the expression, frowned upon.

It should not come as a shock that customers, wittingly or unwittingly, are reading your body language as well. In his book *Honest Signals*, MIT professor Sandy Pentland presented research on the nervous system and "unconscious" communication. He built on the work of others in the area of "mirror neurons." When you watch others move, specific parts of your brain related to the same movement light up. And since "like likes like," other people gesturing the same as you seem to be friendly and sympathetic. At your next reception or networking event, watch two people in agreeable conversation sync up their body language and contrast that with two people not agreeing with each other. If your customer is in a good mood, you might want to consider lightly mirroring their body language. If they are not, you might want to relax and reduce tension in your body so that you don't subconsciously sabotage your messages. For if you don't control your body, your body will control your language.

Dr. Pentland also found a correlation between your body's fluidity/consistency and perception in others of your expertise. The smoother you move, like a golf pro, the more your customer is likely to trust what you say. While you're watching a film, study the body language of the actors. What would they be saying if there was no sound? How does the hero move? The villain? The comic? One of the masters of smooth moves on the silver screen was Cary Grant, who captivated a generation of fans. Why? Because his practiced and fluid body language allowed him to say so much more than words. More recently, I believe, George Clooney and Kevin Spacey are masters of movement. Would you buy what they're selling?

As Lois Lane once said, "Oh, what I wouldn't give for a little x-ray vision."

Vision and Insight Quotes

"The best vision is insight."
--Malcolm S. Forbes

"An artist is not paid for his labor but for his vision."
--James Whistler

"The only thing worse than being blind is having sight but no vision."
--Helen Keller

"Vision is the art of seeing what is invisible to others."
--Jonathan Swift

"In order to carry a positive action we must develop here a positive vision."
--Tenzin Gyatso, the 14th Dalai Lama

"When most I wink, then do my eyes best see."
--William Shakespeare

"If I have seen farther than others, it is because I was standing on the shoulders of giants."
--Isaac Newton

"He who looks through an open window sees fewer things than he who looks through a closed window."
--Charles Baudelaire

"You cannot depend on your eyes when your imagination is out of focus."
--Mark Twain

"The real voyage of discovery consists of not seeking new landscapes but in having new eyes."
--Marcel Proust

"There is nothing so terrible as activity without insight."
--Johann Wolfgang von Goethe

"I have frequently gained my first real insight into the character of parents by studying their children."
--Arthur Conan Doyle

"If one is master of one thing and understands one thing well, one has at the same time, insight into and understanding of many things."
--Vincent Van Gogh

"A moment's insight is sometimes worth a life's experience."
--Oliver Wendell Holmes, Jr.

"A point of view can be a dangerous luxury when substituted for insight and understanding."
--Marshall McLuhan

"There is a condition worse than blindness, and that is, seeing something that isn't there."
--Thomas Hardy

"May you have the hindsight to know where you've been, the foresight to know where you are going, and the insight to know when you have gone too far."
--Irish Blessing

YOUR CALL TO ACTION

Your Call to Action

If anything is harder than being a government contractor, it's being a government customer. Your customers face enormous challenges, so much so that the least difficult option for them to take is the "do nothing option" – to make no decision. When you're working with any customer, your main obstacle to a successful business deal may not be your competition, but a non-decision by the approval authority. They punt the ball until the next fiscal year or next administration or to their successor. In many cases this hurts the customer more than it hurts you. For as common wisdom tells us, doing nothing can be the worst of all possible choices. This is your call to action.

During the first decade of this century in the aerospace and defense sector, business developers could be successful just by establishing friendly relations with government customers and quietly and calmly assisting them with information until decisions were made. That was then. Government customers now are operating with greater constraints and all decisions carry greater risks. As mentioned before, there will be programs that will go forward and there will be decisions that will be made. But will they be made including you?

There's a term entitled "call to action" in sales techniques. It's when you as the salesperson help the customer move from the pain of doing nothing to the better place where your product or service solution is found. It's related to the "burning platform" sense of urgency as you help your customer overcome the fear of taking the leap of decision. You help them understand that staying put can be fatal. This reportedly came from the experience of a Norwegian oil worker who jumped from a fire on a drilling rig to the icy water 150 feet below. When pulled to safety and asked why he did it he replied: "Better probable death than certain death." Here is where customer trust comes into play. In stressful times, no customer will jump where you suggest unless they trust you and see you as a "trusted advisor" who will jump with them.

Trust comes from two important perceptions by the customer: your knowledge and your character. All the time you've patiently been by your customer's side can be enormously helpful in guiding them to a good decision if you play your trusted advisor role and instill an honest sense of urgency – a call to action. Because key acquisition officials have many policies and regulations to consider they may not jump immediately. But with your help they could get ready to. This is what we used to refer to in Texas as the ever-important "Fixin' To" stage. And remember, if you aren't talking now to your customer about where to jump when the time comes, it's likely somebody else is.

There's one last thing to consider. You may have heard the observation that we live out our lives in hero stories. This is well documented by Joseph Campbell in his book *The Hero with a Thousand Faces*, which illustrates how the common hero myth pervades our cultures and compels us to identify with heroes in stories, books, and movies. The trap for business developers and program managers is that we all want to be the hero, to take things into our own hands, and push the customer off the platform to safety. We all want to be Luke or Leia in Star Wars. Instead – be Obi. The message to your customer should be crystal clear: "There's danger. You need to jump for the good of everyone. I'll help show you how and where and I'll jump with you. You be the hero and I'll be with you all the way."

Action Quotes

"I never worry about action, but only inaction."
--Winston Churchill

"Action expresses priorities."
--Mahatma Gandhi

"Action speaks louder than words, but not nearly as often."
--Mark Twain

"The superior man acts before he speaks, and afterwards speaks according to his action."
--Confucius

"Do you want to know who you are? Don't ask. Act! Action will delineate and define you."
--Thomas Jefferson

"Never confuse motion with action."
--Benjamin Franklin

"There are risks and costs to action. But they are far less than the long range risks of comfortable inaction."
--John F. Kennedy

"An ounce of action is worth a ton of theory."
--Ralph Waldo Emerson

"Action is the foundational key to all success."
--Pablo Picasso

"Get action. Seize the moment. Man was never intended to become an oyster."
--Theodore Roosevelt

"I'm a woman of very few words, but lots of action."
--Mae West

"Until the men of action clear out the talkers, we who have social consciences are at the mercy of those who have none."
--George Bernard Shaw

"Our real problem, then, is not our strength today; it is rather the vital necessity of action today to ensure our strength tomorrow."
--Dwight D. Eisenhower

"Action is character."
--F. Scott Fitzgerald

"When I was kidnapped, my parents snapped into action. They rented out my room."
--Woody Allen

"An organization's ability to learn, and translate that learning into action rapidly, is the ultimate competitive advantage."
--Jack Welch

"Nothing is more terrible than to see ignorance in action."
--Johann Wolfgang von Goethe

"Chaotic action is preferable to orderly inaction."
--Will Rogers

"The chief condition on which, life, health and vigor depend on, is action."
--Colin Powell

"Follow effective action with quiet reflection. From the quiet reflection will come even more effective action."
--Peter Drucker

"In action a great heart is the chief qualification. In work, a great head."
--Arthur Schopenhauer

"Action may not bring happiness but there is no happiness without action."
--William James

"To every action there is always opposed an equal reaction."
--Isaac Newton

"Be wary of the man who urges an action in which he himself incurs no risk."
--Seneca

"It is easy to sit up and take notice; what is difficult is getting up and taking action."
-- Honoré de Balzac

"It is most pleasant to commit a just action which is disagreeable to someone whom one does not like."
--Victor Hugo

"Take time to deliberate; but when the time for action arrives, stop thinking and go in."
--Andrew Jackson

"I have always thought the actions of men the best interpreters of their thoughts."
--John Locke

"Between saying and doing many a pair of shoes is worn out."
--Italian Proverb

"Action is eloquence."
--William Shakespeare

"I do not believe in a fate that falls on men however they act; but I do believe in a fate that falls on man unless they act."
--G.K. Chesterton

"Deliberation is a function of the many; action is the function of one."
--Charles de Gaulle

"We cannot do everything at once, but we can do something at once."
--Calvin Coolidge

"Do something. If it works, do more of it. If it doesn't, do something else."
--Franklin D. Roosevelt

"Just Do It."
--Nike

Z

MAKE IT E-Z

Make It E-Z

Government customers, by the nature of their work environment, can become dejected, depressed and demoralized. They have to contend with changing administrations, global conflicts, and economic uncertainties. Is it just me, or are tempers growing shorter every year? Complications that used to seem trivial are now really irritating. The other evening I was setting up a new computer system and became absolutely furious over inadequate explanations of all the variables and options. Why should top dollar equipment be so hard to install and use? Where is the guy who sold it to me and said it was easy?

Upon reflection, I was reminded of that old adage KISS (Keep It Simple, Stupid) we often used in my Air Force days to try and lower the seemingly endless points of possible failure in military operations. Of course, this idea has been around a long time. Leonardo Da Vinci wrote, "Simplicity is the ultimate sophistication." One of Albert Einstein's favorite maxims was: "Everything should be made as simple as possible, but no simpler."

One of my favorite books is *Sustaining Knock Your Socks Off Customer Service* by Tom Connellan and Ron Zemke. You might have seen their popular nine-block chart used to display customer satisfaction data and show the pathway to loyal (repeat buyer) and advocate (selling for you) customers. In their research, Connellan and Zemke found that while "product quality" was important, "ease of doing business" was equally important to customers. This dovetails nicely with Gallup research published in *Human Sigma* showing there really are two aspects to customer satisfaction: technical and emotional. Companies have to satisfy both to keep their customer base and expand their business.

So think about how to reduce the stress load of your customers and *Make It Easy* for them. Instead of adding more process, reduce and simplify procedures. Instead of writing for technical perfection, write to perfectly communicate. Instead of waiting for all the data to come in before answering the customer, tell them what you can now and tell them when they'll get the rest of the story. Instead of just telling the customer what you can't do, tell them what you can do (within legal and contractual requirements). Listen to your customers, find out what their pain is and do something to help make it go away. Anything you can do to simplify the lives of your customers will be appreciated and remembered.

In closing this last section, here are some customer service quotes and some other great quotes I couldn't find space for elsewhere and couldn't part with. I hope you've enjoyed all the quotes and found some to be meaningful and useful. And I hope you've laughed a little, for life is short and happiness is not overrated. You can quote me on it.

E-Z Customer Service Quotes

"Listening is such a simple act. It requires us to be present, and that takes practice, but we don't have to do anything else. We don't have to advise, or coach or sound wise. We just have to be willing to sit there and listen."
--Margaret J. Wheatley

"The only way to entertain some folks is to listen to them."
--Kin Hubbard

"One thing that is often overlooked in leadership is the ability to listen. Listening is so important."
--John Wooden

"Complaining is good for you as long as you're not complaining to the person you're complaining about."
--Lynn Johnston

"If there were nothing wrong in the world, there wouldn't be anything for us to do."
--George Bernard Shaw

"A problem is a chance for you to do your best."
--Duke Ellington

"Those things that hurt, instruct."
--Benjamin Franklin

"A man who has committed a mistake and doesn't correct it is committing another mistake."
--Confucius

"As long as the world is turning and spinning, we're gonna be dizzy and we're gonna make mistakes."
--Mel Brooks

"The man who does not make any mistakes does not usually make anything."
--William Connor Magee

"It is easier to forgive an enemy than to forgive a friend."
--William Blake

"If you haven't forgiven yourself something, how can you forgive others?"
--Dolores Huerta

"When you're angry, never put it in writing. It's like carving your anger in stone. That makes implacable enemies."
--Estee Lauder

"Under certain circumstances, profanity provides a relief denied even to prayer."
--Mark Twain

"There is nothing worse than aggressive stupidity."
--Johann Wolfgang von Goethe

"Only three things happen naturally in organizations: friction, confusion and under performance. Everything else requires leadership."
--Peter Drucker

"An eye for an eye will make the whole world blind."
--Mahatma Gandhi

"Chaos and Order are not enemies, only opposites."
--Richard Garriott

"When a thing ceases to be a subject of controversy, it ceases to be a subject of interest."
--William Hazlitt

"To fly, we have to have resistance."
--Maya Lin

"One of the tests of leadership is the ability to recognize a problem before it becomes an emergency."
--Arnold H. Glasow

"When you have got an elephant by the hind leg, and he is trying to get away, it's best to let him run."
--Abraham Lincoln

"Every problem contains the seeds to its own solution."
--Norman Vincent Peale

"Even if you're on the right track, you'll get run over if you just sit there."
--Will Rogers

"A clever person turns great troubles into little ones and little ones into none at all."
--Chinese proverb

"The best way out is always through."
--Robert Frost

"The secret of managing is to keep the guys who hate you away from the guys who are undecided."
--Casey Stengel

"The most successful people are those who are good at Plan B."
--James Yorke

"Charm is a way of getting the answer yes without asking a clear question."
--Albert Camus

"Take everything you like seriously, except yourselves."
--Rudyard Kipling

"Humor is emotional chaos remembered in tranquility."
--James Thurber

Conclusion

If there are just three actions I hope you'll commit to after reading these sections and quotes, they are these:

Be there when they need you

Be reliable in keeping your promises

Be easy to do business with

These three things will profoundly change your relationship with your customers (and perhaps everyone you come in contact with). Very best wishes on successfully improving customer relations and sales in your favor.

David E. Potts, Ph.D.
President, Tazewell Strategies
Government Business Development Consulting
Customer Relations and Sales Training
604A N. Tazewell St.
Arlington, Virginia 22203
703-850-3061
david@tazewellstrategies.com
www.tazewellstrategies.com

Suggested Reading

Adams, Scott, *The Dilbert Principle*, Harper Collins, 1996.

Ailes, Roger, *You Are the Message*, Currency, 1989.

Ariely, Dan, *Predictably Irrational: The Hidden Forces that Shape our Decisions*, Harper Collins, 2008.

Augustine, Norman R., *Augustine's Laws*, Penguin, 1986.

Axtell, Roger E., *Gestures: The Do's and Taboos of Body Language Around the World*, Wiley, 1991.

Barrett, Thomas, *Dare to Dream and Work to Win: Understanding the Dollars and Sense of Success in Network Marketing*, Business/Life Management, 1998.

Beckwith, Harry, *What Clients Love: A Field Guide to Growing Your Business*, Warner Books, 2003.

Blanchard, Ken, et al, *Customer Mania: It's Never Too Late to Build a Customer-Focused Company*, Free Press, 2004.

Burg, Bob, *Winning Without Intimidation*, Samark Publishing, 1998.

Caruso, David R. and Peter Salovey, *The Emotionally Intelligent Manager: How to Develop and Use the Four Key Emotional Skills of Leadership*, Jossey-Bass, 2004.

Cialdini, Robert B., *Influence: The Psychology of Persuasion*, Collins, 2006.

Collins, James, *Good to Great and the Social Sectors*, Harper Collins, 2005.

Connellan, Thomas K., and Ron Zemke, *Sustaining Knock Your Socks Off Service*, American Management Association, 1993.

Covey, Stephen M.R., *The Speed of Trust: The One Thing That Changes Everything*, Free Press, 2006.

Dixon, Matthew and Brent Adamson, *The Challenger Sale: Taking Control of the Customer Conversation*, Penguin, 2011.

Fast, Julius, *Body Language*, Pocket Books, 1970.

Fleming, John and Jim Asplund, *Human Sigma: Managing the Employee-Customer Encounter*, Gallup Press, 2007.

Fox, Jeffrey J., *How to Become a Rainmaker: The Rules for Getting and Keeping Customers and Clients*, Hyperion, 2000.

Freiberg, Kevin and Jackie, *Nuts: Southwest Airlines' Crazy Recipe for Business and Personal Success*, Broadway Books, 1996.

Gallo, Carmine, *The Presentation Secrets of Steve Jobs*, McGraw-Hill, 2010.

Gladwell, Malcolm, *Blink: The Power of Thinking Without Thinking*, Little, Brown & Co., 2005.

Goleman, Daniel, *Emotional Intelligence: Why It Can Matter More Than IQ*, Bantam, 2006.

Greenleaf, Robert, *Servant Leadership*, Paulist Press, 2002.

Heath, Chip and Dan Heath, *Made to Stick: Why Some Ideas Survive and Others Die*, Random House, 2007.

Hooks, Ivy and Kristin Farry, *Customer-Centered Products*, AMACOM, 2001.

Johnson, Arlene, *Success Mapping*, Emerald, 2009.

Laughlin, Chuck, *Samurai Selling: The Ancient Art of Modern Service*, St. Martin's Griffin, 1994.

Lindstrom, Martin, *Buy-Ology: Truth and Lies About Why We Buy*, Doubleday, 2008.

Mackay, Harvey, *The Mackay MBA of Selling in the Real World*, Penguin, 2011.

Maister, David H., et al, *The Trusted Advisor*, Free Press, 2001.

Morrison, Terri and Wayne A. Conaway, *Kiss, Bow or Shake Hands: The Bestselling Guide to Doing Business in More than 60 Countries*, Adams Media, 2006.

O'Guin, Michael and Kim Kelly, *Winning the Big Ones: How Teams Capture Large Contracts*, O'Guin, 2012.

Page, Rick, *Hope is Not a Strategy: The 6 Keys to Winning the Complex Sale*, McGraw-Hill, 2002.

Patterson, Kerry, et al, *Crucial Conversations: Tools for Talking When Stakes Are High*, McGraw-Hill, 2002.

Pentland, Alex, *Honest Signals: How They Shape Our World*, MIT Press, 2008.

Pink, Daniel H., *To Sell is Human: The Surprising Truth About Moving Others*, Riverhead Books, 2012.

Rackham, Neil, *SPIN Selling: Situation, Problem, Implication, Need-Payoff*, McGraw-Hill, 1998.

Rich, Ben R. and Leo Janos, *Skunk Works*, Little Brown & Co., 1994.

RoAne, Susan, *How to Work a Room: The Ultimate Guide to Social Savvy*, Harper Collins, 2000.

Selden, Larry, *Angel Customers and Demon Customers*, Penguin Books, 2003.

Sewall, Carl and Paul B. Brown, *Customers for Life*, Currency, 2002.

Seybold, Patricia B., *The Customer Revolution*, Crown, 2001.

Shelton, Ken, Editor, *The Best of Class: Building a Customer Service Organization*, Executive Excellence Publishing, 1998.

Thompson, Harvey, *Who Stole My Customer?? Winning Strategies for Creating and Sustaining Customer Loyalty*, Prentice Hall, 2004.

Ury, William, *The Power of a Positive No: Save the Deal, Save the Relationship - And Still Say No*, Bantam Books, 2007.

Willingham, Ron, *Hey, I'm the Customer: Front Line Tips for Providing Superior Customer Service*, Prentice Hall, 1992.

Zander, Rosamund Stone and Ben Zander, *The Art of Possibility: Transforming Professional and Personal Life*, Penguin, 2000.

Customer Relations and Sales Training Partners

John Asher
Chairman and CEO, Asher Strategies
DC Sales Training and Coaching
Sales and Marketing Process Improvement
1300 13th Street NW #608
Washington, DC 20005
866-732-0363
www.asherstrategies.com

Dr. Tom Barrett
President, Business/Life Management, Inc.
Customer Relations Seminars and Online Classes
Keynote Speaker on Leadership and Communication
43543 Butler Place
Leesburg, Virginia 20176
703-963-2813
tom@daretodream.net
www.daretodream.net

Arlene Johnson
President, Sinequanon Group, Inc.
Customer Value Conversation Seminars
Value Based Negotiation Seminars
Miller Heiman Sales Performance Courses
14902 Preston Road, Suite 404-531
Dallas, Texas 75254
972-991-6991
ajohnson@sgroupinc.com
www.SinequanonGroup.com

www.ingramcontent.com/pod-product-compliance
Lightning Source LLC
LaVergne TN
LVHW051039080426
835508LV00019B/1612